Future of North Korea 2032

Coevolutionary Strategy for the Advancement

EAI
EAST ASIA INSTITUTE

East Asia Institute
909 Sampoong Bldg, Euljiro 158
Jung-Gu, Seoul 100-786
Republic of Korea
T. 82-2-2277-1683 F. 82-2-2277-1684
www.eai.or.kr eai@eai.or.kr

Future of North Korea 2032:
Coevolutionary Strategy for the Advancement
Edited by Young-Sun Ha and Dongho Jo

ISBN 978-89-92395-21-2 93340
USD 20.00 / KRW 20,000

June 1, 2012
Printed in Seoul, Republic of Korea

FUTURE OF
NORTH KOREA
2032

Coevolutionary Strategy for the Advancement

EDITED BY YOUNG-SUN HA AND DONGHO JO

PREFACE

BOTH PAST AND CURRENT SOUTH KOREAN GOVERNMENTS HAVE shared a common policy goal toward North Korea to seek change. Presidents Kim Dae-jung and Roh Moo-hyun adopted engagement approaches, whereas the Lee Myung-bak administration opted for a more principled engagement strategy. The engagement policies of the Kim and Roh administrations believed that Pyongyang could be made to change through unconditional support while the principled engagement policy under the Lee administration expected that strong pressure would force the regime to change. Neither policy though considered that change would also have to come from Seoul. As all these policies were nothing more than hopeful efforts, no government could escape from criticism over the ineffective results from these different approaches. In fact any real shift in policy should come from within and not as a result of external forces.

It is now clear how we should deal with Pyongyang. South Korea should pursue a strategy where North Korea chooses to implement change from within. The adoption of a "coevolutionary approach" is essential, utilizing a policy that pursues changes from South Korea and neighboring countries in the region in accordance with changes from North Korea to encourage its momentum to reform.

Throughout the three year long research, the continued support of the East Asia Institute (EAI) has been a driving force toward helping the authors to pursue this project. It could not have been completed were it not for the cooperation of the EAI, from the administrative support including budget management and provision of research space to the publication of this book. We appreciate the generous support of President Sook-Jong Lee and the professionalism of her staff. In particular, we would like to acknowledge the work of Hye-Young Baek of External Affairs Unit, Eun Hae Choi and Ha-jeong Kim of Peace and Security Research Unit, and Young-Hwan Shin of the Publication Unit. We hope that this book will be of benefit to academics and policy practitioners who are concerned with the future of the Korean Peninsula.

Dongho Jo

CONTENTS

CHAPTER 1

Path to an Advanced North Korea by 2032 … **9**
Building a Complex Networked State | ▪ *Young-Sun Ha*

CHAPTER 2

North Korea as a Transformer … **29**
From a Fortress State to an Amphibious State | ▪ *Seongji Woo*

CHAPTER 3

Coevolution Strategy for North Korea … **53**
Diplomacy | ▪ *Chaesung Chun*

CHAPTER 4

Getting Out of the Military-First Dilemmas … **95**
In Search of North Korea's Coevolution Military Strategy | ▪ *Jihwan Hwang*

CHAPTER 5

A Strategy for the Advancement of the … **121**
North Korean Economy | ▪ *Dongho Jo*

CHAPTER 6

A Strategy for the Coevolution of … **145**
North Korean Human Rights | ▪ *Soo-Am Kim*

CHAPTER 1

Path to an Advanced North Korea by 2032
Building a Complex Networked State

Young-Sun Ha

THE NORTH KOREAN LEADERSHIP HAS OFFICIALLY DECLARED THAT NORTH Korea will become a *gangseongdaeguk*, or "strong and prosperous state," by 2012. But the outlook is grim. Despite its efforts to develop a nuclear military-first ideology, North Korea is walking down the path of deepening insecurity and uncertainty, followed by additional excessive security measures. Its currency reform was made to no avail. Excessive spending for security has made it impossible for Pyongyang to invest sufficiently in culture, the environment, and information technology, which are all necessary aspects of developed states in the twenty-first century. Last, Kim Jong-il's dire health condition has led Pyongyang to the inevitable next succession, and it is unclear whether his successor, Kim Jong-un, will persist in military-first politics or redirect the country to an economy-first politics.

In the 1990s North Korea was faced with three hardships. It experienced the fall of communism in 1991, and it had to bear the death of Kim Il-sung in 1994. Then, after going through the Arduous March, a terrible famine that killed a significant number of the total population, the second Kim's regime raised a new banner in 1998 to build a *gangseongdaeguk* as a goal of the state for the twenty-first century. In the same year on May 22, *Rodong Sinmun* (Newspaper of the workers) described the word as defining an "audacious blueprint to lead the country and shine upon the new century." The

Democratic People's Republic of Korea (DPRK) said that "numerous states and people had attempted to achieve a lasting power and prosperity, but history had yet to see a perfect *gangseongdaeguk*." It emphasized that the DPRK would start by building a "nation of strong ideology" with Kim Il-sung at the center, then establish "mighty armed forces" as the pillar for the revolution, and last add a "flourishing economy." "That is Our General's [Kim Jong-il's] own way of *gangseongdaeguk*" (*Rodong Sinmun* May 22, 1998).

Pyongyang designated the year 1999 to be the turning point for *gangseongdaeguk* in the 1999 New Year's Joint Editorial. Now ten years have passed, and it is still suggesting its way of *gangseongdaeguk*, with the three pillars of political ideology, military, and economy, as the survival strategy for the twenty-first century. With political ideology and military already established to some extent, the DPRK is placing emphasis on the economy, which might improve the standard of living as well as help develop heavy industry (*Rodong Sinmun* January 1, 2010).

On June 16, 1999, in a joint article titled "Our Party's (Worker's Party) Military-First Politics Is Invincible," Pyongyang claimed that "the Workers' Party of Korea (WPK)'s policy of giving priority to military affairs is a powerful policy that ensures a decisive victory in long-term confrontation with imperialism." It added that "it is the perfect mode of politics in the present times." Also, the DPRK explained that "military-first politics is a mode of leadership which solves all problems arising in the revolution and construction on the principle of giving top precedence to military affairs and pushes ahead with the socialist cause as a whole, putting forward the army as the pillar of the revolution." According to these writers, military-first politics is a means to strengthen the revolutionary army, guarantee the independent status of the people, and maximize the creative roles of the people (*Rodong Sinmun* June 16, 1999).

The DPRK explained that "changes in circumstances around the *Juche* state" and "unprecedented hardships in the face of the revolution" are the reasons for Kim Jong-il's decision to focus on military-first policy in the mid-1990s. The fall of the communist bloc, the death of Kim Il-sung, the military threat from the United States, and the Kim Young Sam administration's hard-line North Korea policy are the four factors Pyongyang listed as alarming "changes in circumstance." As for the "unprecedented hardships," the following three factors were spelled out: first, the collapse of the communist market, which accounted for 70−80 percent of North Korea's overseas exports, seriously undermined North Korean economy; second, the financial sanctions by the United States and Japan "completely blocked international finance channels" and "strangled North Korea by economic means"; last, "years of natural disasters such as floods, tidal waves, and drought" befell North Korea. As a consequence, shortages of food, fuel, and energy ensued and the Arduous March began. The DPRK records the situation of that time as follows: "History records the 900 days of the Leningrad Blockade as the most catastrophic event. But our whole country, not one city, [has been] surrounded by enemies for much longer than 900 days. These enemies are not one but numerous imperialists attacking us every single day. This is truly unprecedented" (Kang 2002; Y. Kim 2005; Chun 2004; B. Kim 2005).

In order to overcome the Arduous March, the DPRK set January 1, 1995, as the first day of the military-first era, but it was not formalized until 1998. At the first meeting of the Tenth Supreme People's Assembly held on September 5, 1998, Kim Jong-il was promoted to Chairman of the National Defense Commission and the DPRK constitution was revised to establish military-first politics with the National Defense Commission in the center. On that day, Kim Young-nam, who gave the welcoming address, described the role of the Chairman of the National Defense Commission as "the highest post of the state with which to organize and lead the work of defending the state system of the socialist country and the destinies of the people; and strengthening and increasing the defense capabilities of the country and state power as a whole through command over all the political, military and economic forces of the country. It is also a sacred, important post which symbolizes and represents the honor of our country and the dignity of the nation" (*Rodong Sinmun* September 7, 1998).

Decline of Military-First Politics

Military-first politics, which was chosen as a new political means to build twenty-first-century *gangseongdaeguk*, brought about the exact opposite of what was intended: a vicious circle of failing security, economy, and politics.

Vicious Circle of the North Korean Nuclear Issue

The primary goal of military-first policy was protection of the Kim Jong-il *Suryong,* or Great Leader Direct Rule[1] system, at all costs when the socialists' international order collapsed. In the "Agreed Framework between the United States of America and the Democratic People's Republic of Korea in 1994," the United States proposed construction of a light-water reactor and provision for heavy-water reactor in return for a complete freeze on the North Korean nuclear program as well as the disassembling of existing nuclear facilities. The United States also proceeded with negotiations to improve the relationship with the North, but the DPRK did not abandon developing nuclear weapons because it considered them the last hope for the security of the system. The efforts of the Clinton administration made it possible to hold a Washington meeting with North Korea, but due to Pyongyang's uncompromising demands for excessive security, the United States failed to achieve tangible results. Furthermore, after 9/11, the United States began to treat nuclear proliferation and terrorism not merely as a matter of global security but as a matter of highest national security. This implies that North Korea's nuclear program became a matter of life and death for both Washington and Pyongyang.

The second DPRK nuclear crisis began when a high-enriched uranium (HEU) program was brought to the table by the U.S. deputy minister of the Asia-Pacific James Kelly, who visited Pyongyang in early October 2002. In the end, the Six-Party Talks, which began with Beijing's mediation, tentatively agreed on the September 19 Joint Statement in 2005. But the core four ele-

1) [Translator's Note] *Suryong* holds supreme power over decision making in North Korea (K. Kim 2008, 90).

ments of this Joint Statement, "North Korea's nuclear abandonment," "financial support," "normalization of relations," and "peace system," faced the same fate that befell the "Agreed Framework between the U.S. and the DPRK in 1994": no practical outcome. The DPRK demanded a lifting of economic sanctions, financial support, establishment of a peace system, recognition of the North Korean regime, and improvement in U.S.-DPRK relations as preconditions for abandoning its nuclear program. On the other hand, the United States emphasized verifiable nuclear abandonment prior to any progress in other areas (Ha 2006).

As expected, the Six-Party Talks failed to reach any agreement due to financial sanctions against North Korea, involving Banco Delta Asia (BDA) in Macao, and Pyongyang launched a nuclear test in October 2006. In 2007, both sides agreed to the three fundamental steps to fulfill the Joint Statement —disable the facilities, report the progress, and finally denuclearize—but they failed to narrow the gap in the steps for report and verification. North Korea soon declared the annulment of nuclear negotiations in April 2008 and restored its measures of freeze and dismantlement to its formal conditions. On May 25, 2009, it launched a second nuclear test. The North Korean nuclear issue repeated the vicious circle of economic sanctions and nuclear tests with no viable solution. In early October 2009, Chairman Kim Jong-il met with Chinese prime minister Wen Jiabao and announced: first, "denuclearization of the Korean Peninsula" was the will of his deceased father, Kim Il-sung; second, the hostile relations between North Korea and the United States must be improved through bilateral talks between the two; and last, the DPRK intended to participate in the multilateral talks, including the Six-Party Talks, depending on the results of bilateral talks with the United States (*Rodong Sinmun* October 6, 2009). In early May 2010, Kim Jong-il repeated these conditions during his visit to Beijing (*Rodong Sinmun* May 8, 2010).

However, the fall of nuclear military-first politics is only a matter of time. Nuclear weapons of the *Suryong* system provide a warrant not for life but for death. If the North keeps rejecting the strategic decision to abandon the nuclear program, the U.S.-led economic sanctions will be intensified, and the possibility that the United States and China would implicitly allow an emergence of a non-nuclear, pro-Chinese regime in the North will mount. After 9/11, the United States has been placing top priority on eliminating weapons of mass destruction and terrorism. North Korea would be sorely mistaken to expect the United States to allow its nuclear program in any case.

Failing in Economic Growth

The Kim Jong-il regime has put the banner of a *gangseongdaeguk* for its twenty-first-century nation building and emphasized military-first politics and ideology, military, and economy as the main focus of its efforts. While all of those are essential, the relative importance differs. Military-first politics is the top priority, followed closely by a strong military, with economy last. Therefore, when the logic of the economy has clashed with that of military-first politics, the latter has always taken precedence (Ha 2000).

In the midst of deepening crisis, North Korea chose a military-first economic strategy to prioritize the defense industry over light industry and agriculture, even during the Arduous March. This decision can be seen as a desperate action taken at the brink of collapse after the fall of the communist bloc.

The military-first economy attempted a twenty-first-century version of *wijŏng ch'ŏksa*, or defense of Confucian orthodoxy and rejection of Christian heterodoxy[2)]—that is, a rigid way of looking at its own future in a dichotomy of self-reliance and subordination—outwardly, and domestically allocated its meager capital into heavy industry and defense, all of which failed to resolve the imminent financial crisis. In the end, Pyongyang carried out the "July 1 Measure" in 2002 and implemented "*jonghabsijang*," or the "Comprehensive Market System," in 2003. But even these efforts were carried out within the limits of the military-first economy, and North Korea's economy went through negative growth for ten years consecutively from 1990. Despite slight growth in the 2000s, North Korea's GDP remains at 500 to 1,000 dollars per capita, one of the lowest in the world.

In a new attempt, Pyongyang carried out a currency reform in November 2009, and since 2010, it has been emphasizing the importance of light industry and agriculture. However, if it does not make a sweeping decision to emerge from the limits placed by military-first politics, the DPRK will inevitably fall deeper into financial crisis.

Establishing a System for Succession

Kim Jong-il, who inherited the regime after Kim Il-sung's death in 1994, initiated "rule by his deceased father's will" in the face of the deepening crisis regarding the domestic economy and security instability derived from the collapse of communism. As soon as he seized power, Kim Jong-il established a core group with his blood relatives in key posts, whose ultimate purpose was the survival of Kim's family. First, Kim and his followers maneuvered a policy of military-first politics, which prioritized the protection of the *Suryong* system as the foundation for all organizations and activities of the Korean Workers' Party, through the Organization Guidance Department (Administration Council Director Chang Sung-taek) of the Central Committee Secretariat and Propaganda and Agitation Department. Kim then took control of the military with the Korean People's Army General Political Department (Senior Deputy Director Kim Jong-gak), General Staff Department (Chief Yi Yong Ho and Operations Bureau's director Kim Myong-kuk), Guard Command (Commander Yun Chong-nin), Pyongyang Defense Command, and Military Security Command (Chief Kim Won-hong) at the core of the power structure. Last, Kim and his relatives needed working-level bureaucrats for the cabinet to resolve the food crisis of 20 million people in North Korea.[3)]

In mid-August 2008, Kim Jong-il's health deteriorated, rendering the succession issue urgent, which would be the second power succession in the sixty years of North Korean history. Compared to the first succession (from Kim Il-sung to Kim Jong-il) this recent power transition has not allowed allow sufficient time to prepare. The issue at hand is whether this succession will remain in the "rule by his deceased father's will," or begin to seek a new survival strategy for the new century. It will depend on how Kim Jong-un and

---

2) [Translator's Note] Concerning this term, see Chung (1995) and Kum (2001).

3) Information Center on North Korea (http://unibook.unikorea.go.kr/MA/).

his clique manage to become a new political force under the newly appointed leader.

The First Step of Coevolution Strategy for North Korea's Advancement: Transition and Reform

Efforts of North Korea for Coevolution

Without resolving its three major problems (the vicious circle of the North Korean nuclear issue, the dire economic crisis, and the establishment of a succession system), Pyongyang has had to carry out an incomplete succession due to Kim Jong-il's poor health. In the process of building a succession system, the pace of the change is crucial.A tragic fate befell the Soviet Union because it pushed its reforms too rapidly; on the other hand, China is enjoying the fruits of reform because it succeeded in pacing. The fate of North Korea will largely depend on the relevant pacing with changes occurring in and out of the state.

The first and foremost task for the new regime is survival, guaranteeing the life and existence of the system. To achieve this, North Korea's military-first politics should be reexamined. Fifteen years of military-first politics have been a "lost fifteen years" and Pyongyang has insisted that a nuclear weapon is essential for its survival due to the constant threats coming from the United States. However, the nuclear policy of military-first politics brought about the exact opposite of what it intended: a warrant for the death of Kim's system involving the vicious circle of nuclear tests and economic sanctions. Therefore, for North Korea, to resolve its nuclear problem is the most pressing task before planning a survival strategy for the new century.

The Six-Party Talks, a meeting to find a solution to the North's nuclear issue, passed the first step (disablement) and is fighting an uphill battle at the second step (report and verification), with the third step (denuclearization) as its ultimate goal. These steps cannot be achieved with simple technical agreement but require beforehand the new regime of North Korea to make a strategic decision for survival without nuclear weapons. This strategy should mandate the replacement of nuclear weapons with traditional weapons as the basis of the DPRK defense system, along with the establishment of a peace system for the Korean Peninsula. For far too long, the DPRK has emphasized the peace agreement with the United States as a key to the peace system on the Korean Peninsula. Also, it has stressed that the North Korea-U.S. military committee should deal with the withdrawal of the U.S. forces in Korea and the issue of South Korea-U.S. alliance as a proof of transition toward peaceful relations between the two states.

However, it is necessary to design a new peace system that will be more realistic and acceptable to all of the neighboring countries, including the Republic of Korea(ROK) and the United States. First, instead of nuclear weapons, a comprehensive peace system should be established to guarantee the security of the North's system, within the Six-Party Talks. Previous mistakes that derived from the unrealistic peace agreement must not be repeated, and the utmost efforts are called for to prepare comprehensive security ensurance for the system. Therefore, not only the peace agreement between North Korea and the United States and the one between the two Koreas, but also support from China as well as the multichanneled security system such as the Six-Party Talks are critical. Second, issues on the United States Forces Korea(USFK) and the ROK-U.S. military alliance can be reevaluated in a new light when North Korea makes the necessary strategic decisions and inter-Korean relations begin to steer away from mutual hostility. The denuclearization of the North will expedite its relations with the United States and Japan and reinvigorate financial support for North Korea. Meanwhile, North Korea should reevaluate China and Russia at a new level. China will eventually prefer that North Korea gradually turn into a denuclearized economy-first regime sustained by Chinese influence. In the end, Pyongyang needs to build a sophisticated comprehensive network to simultaneously engage the United States, Japan, China, and Russia. North Korea should no longer linger on its self-reliance motto.

The new survival strategy for North Korea's advancement is to make strategic decisions on denuclearization and its peace system and to promote a transition from military-first to economy-first politics. As shown in China's case in the late 1970s since its reform, North Korea should prioritize economic development. As the latest starter in the market economy, what North Korea needs is not a limited adoption of the market system within the frame of its planned economy but, instead, a transitional reform that combines the two systems.

To successfully realize the above-mentioned reforms, three steps should be taken simultaneously. First, domestically, Pyongyang must carefully implement a transition from a planned economy to a market economy. Then, it must get away from its worn-out method of "*Minjok Kongjo,*" or "exclusive cooperation of radical nationalist," and instead pursue economic development in cooperation with reformists in South Korea — not the ones seeking the status quo of the North. Last, North Korea must overcome the outdated distinction between socialism and capitalism, and should actively make use of the United States, Japan, and China, regardless of such distinctions.

Efforts of South Korea and the International Community for Coevolution

The first step for North Korea's difficult journey to become an advanced state is to pursue a new survival strategy that extends beyond military-first politics. But this grand project can hardly be achieved by North Korea alone. Coevolution strategy involving South Korea and other neighboring states is vital (Garnsey and McGlade 2006). The international community has thus far sought to make changes in North Korea in two main ways: sanctions and the sunshine policy. Unfortunately, neither of those succeeded in bringing change to North Korea. The advocates of the sunshine policy argue that frequent diversions in policies did not allow enough time for any significant results. Though the sunshine policy helped to extend inter-Korean exchanges, it failed to stop North Korea from both developing nuclear weapons and pursuing military-first politics. But at the same time, undiscriminating sanctions do not work as a perfect solution. The system of North Korea is designed to weather such sanctions longer than any other state in the world. It is thus extremely difficult to bring down military-first politics through sanctions. Instead, what is needed are simultaneous changes occurring inside and outside North Korea. The sunshine policy, however, makes a too-rosy assumption that someday and somehow the sunshine will penetrate to politics and the military without practically assessing the domestic difficulties of North Korea. On the other hand, supporters of sanctions put all the responsibilities on the North's systemic attributes. In short, neither of those policies is likely to succeed unless they work in tandem for coevolution.

When Kim Jong-un's regime adopts a post-military-first politics system, it will be immensely important for the related parties to upgrade their ideas on security and prosperity to fit the new system. Currently, North Korea argues that its nuclear program is a countermeasure against hostile U.S. policies toward itself, emphasizing that it can forgo the nuclear program if the United States retracts its hostile policies. This clearly is an excessive security stance of military-first politics. If the Kim Jong-un regime promotes an appropriate level of security under the new system, South Korea and the United States should be able to convince the North that their North Korea policies are not hostile toward the new regime. This can only be achieved through a three-level comprehensive peace system that extends beyond the Korean Peninsula to include all of East Asia. This system will provide the North's new regime with a sense of security that Kim Jong-un's father's worn-out tactics of nuclear threat cannot give. The first step for a three-level comprehensive peace system is to establish the inter-Korean peace system; second, an East Asian peace system under the guarantee of the United States and China; last, a multichanneled peace system such as the Six-Party Talks.

In addition, international communities should actively provide North Korea assistance to succeed in its new economy-first politics. To successfully implement economy-first politics, North Korea should, above all, let go of the fear that capitalism will bring down the North Korean economy. Hence, it is necessary to reexamine why back in the mid-1990s the newly emerged leader, Kim Jong-il, made the fatal mistake of choosing military-first politics as his father willed. In 2012, it ought to be different.

The Second Step of Coevolution Strategy for North Korea's Advancement: Transformation

Norms of Twenty-first-century Civilization

For North Korea to survive in the twenty-first century, a decade of the first step of coevolution strategy is not sufficient. Another ten years of change for the second step of coevolution is needed.[4)]

The second step requires North Korea to meet the norms of twenty-first-century civilization. Following this norm, the related parties should promote to advance North Korea, unify the Korean Peninsula, and implement a coevolution strategy involving East Asia and the world. The new model can be described metaphorically as "wolf spider (Actor) building Dabotap (Stage)" and more analytically, the theory of a complex network. Unlike other spiders, the wolf spider does not weave a web to capture its prey; rather, it attacks them like a wolf does. This metaphor precisely illustrates the necessary abilities required for the leaders of the world order today: a vital combination of two different functions of a wolf and a spider.[5)] The East Asian order of the twenty-first century forms a complex order that shows a modern balance of

Figure 1. Simple Web and Complex Web

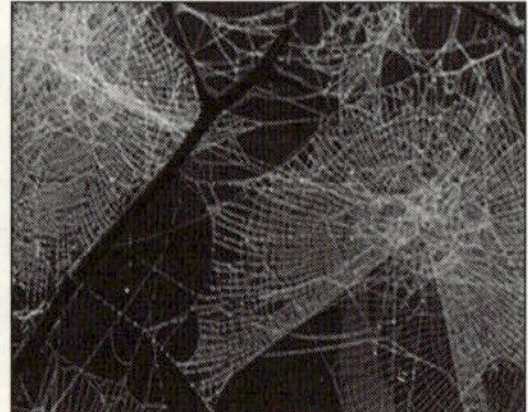

4) This part is a modified version of statements presented at the "Future Strategies Forum" by Korea Institute for Future Strategies (Ha 2010).

5) [Translator's Note] This indicates that the twenty-first-century actor should be able to act as a wolf (who performs self-help or practices a balance of power policy, which is a modern international political principle of Hobbes) and a spider (who establishes network flexibly among diversified actors, which should be a basic principle in the postmodern world) at the same time.

power and, at the same time, the two layers of the spiderwebs that are largely drawn by the United States and China. But it should not be ignored that smaller states are also forming complex diplomatic webs of their own in the midst of the power struggle between the two superpowers. In Figure 1, the picture on the left shows the spiderweb that is normally seen, and the picture in the middle is a two-layered spiderweb, visualizing the rough sketch of bipolarized East Asia. The picture on the right is an example of complex webs, a metaphor for the complex East Asian network.

Let us explore the complex stage of world politics through the metaphor of Dabotap. While security and prosperity have been the traditional stage of the nineteenth century, the twenty-first century introduced an additional set of stages: environment and culture — newly emerged stages as the world suffers

Figure 2. Dabotap: Three-Story Stone Pagoda

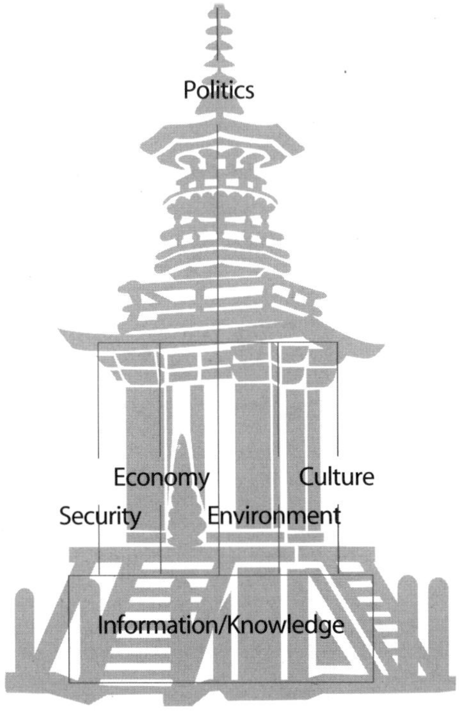

from indiscreet developments and the identity gaps among main actors of the world as they diverge farther apart since the end of the Cold War. As shown in Figure 2, environment and culture, along with traditional stages of security and economy, are the four pillars of Dabotap. At the foundation of Dabotap is the information and knowledge of today. The reason why not economy but information and knowledge are the substratum can be found in history. In ancient times religions were the bases of most people's lives, and in the era of political revolution, politics became the fundamental of lives; after the Industrial Revolution, economy played the role of substratum. However, in the twenty-first century, with revolutionary progress in science and technology, information and knowledge began to influence the four main stages as an independent variable, consolidating its role as a foundation. Politics, on the other hand, still keeps its place at the top of Dabotap. No state can stand at the center of the global stage without building all three layers of the complex system we see in Dabotap in the new century.

South Korea is moving toward a global stage, and it will play a new role in forming the complex network with other major actors of the world. China is on a similar path. For example, in 2008 at the thirtieth anniversary of China's economic reform, President Hu Jintao suggested China's future image for the year 2049 (the hundredth anniversary of the new China).[6] The president stated that the last thirty years would witness the dramatic increase in GDP per capita to more than 3,000 dollars and by 2021, the hundredth anniversary of the foundation of the Chinese Communist Party, China is targeting a GDP per capita of 10,000 dollars. By 2049, Beijing will establish a prosperous, democratic, civilized, and harmonious socialist state. But this overly long title without any concrete explanation implies that it is not very successful in suggesting an attractive Chinese model for the twenty-first-century world. In this regard, experts in the United States and Europe are already

---

6) "中共中央總書記胡錦濤在紀念改革開放30周年大會講話." December 18, 2008 (http://www.beijingreview.com.cn/document/2009-05/11/content_195251.htm). English version: Hu Jintao's speech at the meeting "Marking the 30th Anniversary of Reform and Opening Up" (http://www.china.org.cn/archive/2009-05/11/content_17753659.htm).

at the table discussing the future of China. They have categorized its future into three groups: succeed, fail, or crisis. Still, regardless of the success of the plan, it is important to keep in mind that the direction of development should go through the stage of strength and wealth, as China's case shows, before it enters the next stage, where the actor plays out the performance of a complex network.

China's 2049 model can serve as a guideline for North Korea. Each of the aforementioned concepts—"wolf spider," "Dabotap," and building an advanced North Korea through the coevolution strategy—will be a tough yet inevitable journey for Pyongyang. If the Kim Jong-un regime insists on military-first politics for the thirty years to come, it will undoubtedly face the second and third Arduous Marches, and become a vegetative state in the end. But if the new regime makes a decision to follow the path of China, initiates the first step of advancement (transition and reform), carries out a sweeping economic reform, and subsequently takes the second step of advancement (transformation) and forms a complex network state, reunification would no longer be a blind hope. Yet such a reform can only be achieved through a coevolution with the international community, including South Korea, the United States, and China.

The advancement of North Korea must move on to the second step of building a complex network from the enhanced conditions that it built during the years in the first step. In order to do so, North Korea should first overcome its paltry efforts to sustain the system and make ends meet and move up to the whole next level of establishing an appropriately secure and prosperous state. Meanwhile, it must not neglect environment and culture in order to complete the four main pillars of Dabotap. The next step would be to renew the information and knowledge foundation. In this step, North Korean leaders should transform the oversized military technology under military-first politics into a peaceful technology. As the last step, politics, the top of Dabotap, should steer away from military-first politics, pass the transition period of advanced *Suryong* leadership, and ultimately reach democracy in the end.

Along with the advancement of North Korea, the two Koreas should transcend the stage of modern contacts and be able to meet in a new, comprehensive way. Not as a state-to-state relation, but involving various actors in a comprehensive manner. The modern idea of unification is a matter between the related states, and therefore politics, the military, and the economy are primary concerns. However, in order to develop the inter-Korean relationship within the frame of a complex network, a far more complicated network must be built. A complete reconstruction of all three stories of Dabotap must be deliberately pursued. Simultaneous building of Dabotap for Korea, East Asia, and the world must be carried out while they help North Korea to develop its own. Eventually, the two Koreas will form comprehensive states of the twenty-first century.

Complex Network Reunification and East Asia

Should the way of Korean reunification be achieved via complex network, the East Asian order, which must also work toward coevolution, is a matter of significance as well. The East Asian community will not arrive anytime soon. It will become a reality only when individual states in East Asia and East Asia itself comprehensively solidify their identities, because in principle, it requires heart-to-heart contacts. For the future of the East Asian community, states in this region will have to form dual identities in a long-run process. That is, for North and South Korea a threefold identity is needed for reunification: identity of North or South Korea, of the Korean Peninsula, and of East Asia. Yet the reality lags far behind this idea, as the East Asian community is currently being developed in limited areas only. Considering that the formation of community is rendered possible only through aggregating mutual interests and values, it can be said that East Asia is at its beginning stage. It has just started to build a complex network that will lead its way to the East Asian community and, in a broader sense, the international community.

The most important task remaining for the Korean Peninsula is to prepare a survival strategy in the midst of the new East Asian order led by the two superpowers. For this, the "politics of the affair," it is inevitable that South Korea should embrace the "secondary wife," China, while maintaining its relationship with the "first wives," the United States and Japan. There is no way to survive for small powers when reflecting personal ethics to international politics. The "politics of the affair" is, in terms of international politics, politics of the complex network. In the process of reunification, the most important web for the advanced two Koreas is the "dense network," neighboring ones, and the thickest web should be formed with the United States, along with the one with Japan as a sidekick. The alliance with the United States began during the Cold War era in the 1950s, primarily with military cooperation for survival. Since the new century, it has developed from a strategic alliance toward a complex network alliance. A "connected network" with China is are less important for South Korea. ROK-China relations, currently delineated as strategic cooperation, are less than twenty years old and the two states are based on quite different economic and political systems. Yet the swelling trade volume between the two has already surpassed that of ROK-U.S., rendering mutual dependency ever more inevitable (Burt 2007; 2010).

During the Cold War era, when the United States and China's relations were hostile, it was impossible for South Korea to maintain friendly relations with both at the same time. But now, in the new century, ROK-U.S. and ROK-China are not necessarily mutually exclusive. The complex time of the twenty-first century calls for complex networks. As of now, the United States fully understands that China is an asymmetric G2 state, and their relationship cannot remain hostile given China's geopolitical and geoeconomic importance to South Korea. The United States is just hoping to stay as the biggest ally in South Korea's twenty-first-century complex network. Therefore, in the current state of affairs, it may be possible for South Korea to weave a web toward China while sustaining mutual trust with the United States. What is of concern, rather, is China's "Project 2049." China now focuses on peaceful economic development, but as its economy swells in coming years, the theory of the "Chinese standard of civilization,"or Sinocentrism, may loom large on the horizon. Before that happens, China and neighboring states have to put their best efforts into settling China into a proper network of East Asia. With all these tasks that await, the next thirty years of East Asia will be the "busy thirty years."

The United States and China entered a period of complex relations in this new century, overcoming the hostilities from the Cold War era. The Korean Peninsula must strengthen its alliance with the United States, weakening but still a leading state, and develop a future-oriented network with Japan, while transforming the relationship with China into the twenty-first-century version of a new alliance. If the G2 order is solidified without a proper network between the United States and China, South Korea would face serious difficulties. South Korea must embrace both the United States and China, though with different approaches, so it is important that these two webs should not clash with each other.

Weaving a Complex Web on the Korean Peninsula

To complete the unification based on a complex network within the two-step advancement model for North Korea, a new view on the unification for the twenty-first century is required. The key to this view is overcoming the relative disadvantages in resource power with advantages in network power. Recent research on business suggests the following tips to build stronger network power. First, weave the "dense network" as densely as possible. Applying this to the issue at hand, South Korea will need to develop the existing ROK-U.S. alliance into a more complex alliance while strengthening the ROK-Japan relationship. Second, expand the "connected network" as broadly as possible. It is particularly important that South Korea furthers its existing network with the United States and Japan to include China into the complex network. In addition, South Korea's network web must also reach to the global level as well as cyberspace. Last, apart from creating a dense and broad web, South Korea should form a knowledge-mediating web that could fill in "structural holes" in the network. When there is a tear in a network web, the role of the mediator — that is, to mend it with information and knowledge — is crucial. South Korea needs to strengthen its network power through playing the role of a knowledge mediator, because DPRK-Japan and DPRK-U.S. relations contain a number of structural holes. Through this three-step reinforcement of a network web, South Korea should be able to complement the relative disadvantage in resource power. South Korea does not have the resource power paramount to that of the United States, China, Russia, or Japan. Therefore, in order to stand side by side with stronger neighboring states in the twenty-first century, the Korean Peninsula of advanced South and North Korea need to be equipped with more sophisticated network power.

Establishing a complex networked state, which is the second step of coevolution strategy for North Korea's advancement, would be possible through the strengthening of the three network powers mentioned above. These are taking place on the Korean Peninsula, in East Asia, the world as well as in cyberspace. Of course, strengthening network power itself would not allow reunification in a traditional style, but it will lead to forming a world network web for the twenty-first century. There have been cases when a state with weak resource power was at the center of the world. For example, the Netherlands had the most sophisticated ability to network in the seventeenth-century world order; thus, it was able to become the dominant power of that time. Unfortunately, the Korean Peninsula has not had the time or energy to think about its network comprehensively due to the divisions between North and South as well as left and right, but it is not too late. If the North and South achieve advancement and reunification through the complex network web extending over East Asia, the world, and cyberspace and succeed in coevolution, the Korean Peninsula's twenty-first-century complex network web will be able to show the world a new standard of civilization. ▪

■ **References**

Burt, Ronald S. 2007. *Brokerage and Closure: An Introduction to Social Capital*. Oxford: Oxford University Press.

———. 2010. *NeighbourNetworks: Competitive Advantage Local and Personal.* Oxford: Oxford University Press.

Chun, Duk-Sung. 2004. *Seon-gun-jeong-chi-e daehan i-hae* [Understanding of Military-first Politics]. Pyongyang: Pyongyang Press.

Chung, Chai-Sik. 1995. *A Korean Confucian Encounter with the Modern World.* Berkeley: Institute of East Asian Studies, University of California, Berkeley.

Garnsey, Elizabeth and James McGlade, eds. 2006. *Complexity and Co-Evolution: Continuity and Change in Socio-Economic Systems.* Chaltenham, UK: Edward Elgar.

Ha, Young-Sun. 2000. "Bug-han haeg-mi-sa-il ju-gi-jeog wi-gi" [Periodic Threat of North Korea's Nuclear Weapons and Missile]. *Gyegan sasang.* Summer: 163-193.

———. 2006. *Bug-hae wi-gi-wa han-ban-do pyeonghwa* [Nuclear Crisis in North Korea and Peace in the Korean Peninsula]. Seoul: East Asia Institute.

———. 2010. "*Han-ban-do bog-hab-geu-mul-mang-tong-il-gwa dong-a-si-a* [East Asia and Korean Reunification through a Complex Network]. Presented at Future Strategies Forum at the Korea Institute for Future Strategies, July. Seoul. http://www.kifs.org/contents/sub3/forum.php?method=info&searchKey=&searchWord=&offset=&sId=2427.

Kang, Sung-Gil. 2002. *Seon-gun-si-dae-ui jo-gug-eul ga-da* [Going to the Homeland of the Military-first Era]. Pyongyang: Pyongyang Press.

Kim, Bong-Ho. 2005. *Seon-gun-eu-lo wi-lyeog-tteol-chi-neun ganggug* [Strong Nation through Military-first]. Pyongyang: Pyongyang Press.

Kim, Kap-sik. 2008. "Suryong's Direct Rule and the Political Regime in North Korea under Kim Jong-il," *Asian Perspective* 32, 3: 90.

Kim, Young. 2005. *Seon-gun-jeong-chi-wa jo-gug-tong-il* [Military-first Politics and National Unification]. Pyongyang: Pyongyang Press.

Kum, Changt'ae. 2001. *Hwa-sŏ-hak-p'aŭi Ch'ŏl-hak kwa Si-dae ŭi-sik* [Philosophy of the Hwasŏ School and the Awareness of His Time]. Seoul: T'aehaksa.

CHAPTER 2

North Korea as a Transformer

From a Fortress State to an Amphibious State

Seongji Woo

THE DIVISION OF THE KOREAN NATION INTO TWO SEPARATE POLITICAL entities that came about with the defeat of Japanese imperialism continues after more than sixty years since it began. South Korea, or the Republic of Korea (ROK), has aligned with the United States and vigorously pursued a strategy of export-oriented development, which has resulted in remarkable economic growth and in turn laid the seeds for gradual democratization. In this double achievement both economically and democratically, South Korea seeks to achieve developed-nation status.

Meanwhile, North Korea, or the Democratic People's Republic of Korea (DPRK), has allied with two continental powers, China and Russia, and has endeavored to build an autonomous planned economy with as little intervention from the outside as possible. Pyongyang's self-imposed isolation under the slogan of *Juche* (self-reliance) has brought about economic stagnation and persistent backwardness. In the era of globalization and interdependence, the Kim Jong-il regime is making little effort to remedy its malfunctioning economic system and, rather, is focusing on protecting its regime security through the development of nuclear weapons and long-range missiles. It will be difficult for the DPRK to escape its status as a fragile state so long as it sticks to its military-first politics (*Songun* politics) and military-first economy.

In the early twenty-first century, the world is facing formidable challenges in the fields of human rights, environmental protection, climate change, the North-South divide, and so on. As a key member of the G20 Summit, South Korea is playing an active role in fighting these problems and finding solutions to them. Seoul is beginning to transform itself from a rule-taker to a rule-setter as it has assumed a role as a link between the advanced states and the developing states on the global stage. It is contributing to the creation of a new global economic order.

With the collapse of communism in the early 1990s, states are competing in a race for peace and prosperity. North Korea, however, is one of the few exceptions in this race. It is doubtful whether North Korea will be able to achieve the goals of democracy, opening, growth, and equality along with South Korea if it continues with the grand strategy of military first and nuclear armament. Yet it is urgent that North Korea struggle hard to escape from its stagnant identity as a fortress state and transform itself into a state that can capture the merits of both sea powers and land powers and become a truly "amphibious" power.

North Korea under Kim Jong-il has shown some hints of change as it sought reconciliation with South Korea and dialogue with the United States and Japan with the goal of eventual diplomatic normalization, and it adopted the July 1 Economic Restructuring Measures in the early 2000s (Park 2002; Shin 2002). Yet these efforts have fallen short of satisfactory political, economic, social, and diplomatic change. The basic problem lies in the calculation of the current leadership, which views reform and opening as detrimental to the core interests of Pyongyang's ruling circle. As long as North Korea sticks to its inefficient planned economy and strengthens its authoritarian rule over the people, it will not be able to reach the paths to either democracy or prosperity.

What would be necessary for North Korea to escape its isolation and backwardness and join the ranks of advanced states? There are two types of advancement, one a process and the other a final stage of development. As a process, the term refers to efforts to adopt and implement global standards; as a final goal, advancement means a situation where the state itself becomes a creator of global standards. Therefore, North Korea's advancement can be summarized as the sum total of Pyongyang's proactive efforts to narrow the gap with extant global standards and to create its own path-breaking standards.

My purpose here is not to predict a most likely scenario for North Korea's future under today's conditions but to seek a gradual course for North Korea's transformation on the assumption that its development into an advanced state is a necessity. This essay is thus a policy proposal to the current and future ruling elites of North Korea for changes that would result in a series of incremental steps toward advancement. Once the ruling circle in Pyongyang comes to the conclusion that military first is no longer sustainable and the normalization of state governance is necessary, and decides to change its grand strategy, what is a possible roadmap for that transformation? What is to be done to reach the lofty goal of advanced status? What steps and programs are necessary for Pyongyang to enter the ranks of advanced states? How can it prevent a tragic collapse of the state in the form of either an implosion or an explosion? I suggest a less painful and less dramatic incremental reform process comprised of three stages: a first stage (the short term), a second stage (the medium term), and a third stage (the long term).

The first stage is the period of decay when the regime's internal and external contradictions deepen as the policies of military-first politics persist. The military-first regime will be unable to find an exit for advancement and democracy. The second stage will be a transitional period when a new leadership emerges with the aim to get rid of the military-first system. In this period, North Korea will opt for a partial reform and opening. In the third and final stage, North Korea's unique *Suryong* (Leader) system of governance will finally come to an end, and a tantalizing balance between the conservatives and the liberals, which will characterize stage 2, will be resolved by the victory of the latter. North Korea will enter a stage of fundamental transformation, following a motto of "wealth first" and "the people first." The level of Pyongyang's competitiveness will rise, and a state that can achieve goals of innovation for its population, society, and institutions will emerge.

The Evolution of North Korea as a Fortress State

The Fortress State

A state can assume the benign characteristics of a provider or a protector, or it can take the villainous role of a predator attacking its own people. How should we characterize the nature of North Korea as it now stands? The notion of a fragile or failed state is helpful in characterizing, or making sense of, the North.

Fragile states are those that are incapable of performing their necessary functions, and under these states people's lives are heavily constrained. More specifically, such states do not (1) protect their people from violence, (2) provide the public goods that satisfy the basic needs of their people, or (3) possess a representative and unified government within a given territory (Lee C. 2008). Similarly, a failed state refers to a situation where the basic requirements for the provision of welfare, defense of national boundaries, and representation of the people are not met. Collapsed states and fragmented states are extreme cases (Chesterman, Ignatieff, and Thakur 2005; Vinci 2008).

North Korea can be categorized as both a failing state, if not a failed one, and a fragile state. North Korea's governance is not successful in that it fails to deliver goods and service that are basic to its people's basic lives. However, North Korea stands apart from other fragile states in that it is at once fragile but also able to sustain its monopoly on power and is seen as threatening by its neighbors. The duality of economic frailty and military resilience gives North Korea a special status.

The concept of a fortress state is a symbolic notion that captures the characteristics of North Korea. The fortress state is defined as one that is isolated from the international community, considers regime security its top priority, is controlled by a single leader who is free of intervention by other actors or institutional constraints, and continually mobilizes its people for the accomplishment of state goals. In contrast, South Korea is dubbed a voyager state, which has sought to keep a pluralistic society and amplify interdependence with foreign markets through an export-oriented developmental strategy (Woo 2008).

The origins of Pyongyang's self-imposed isolation go back to the daunting experience of the Korean War (1950–1953), in which it had to face the military prowess of the United States, the superpower of the Cold War era. North Korea perceived that it was surrounded by great powers whose interests seldom coincided with those of minor powers. The fortress state has in part been made possible by topographic features: the Yellow Sea, the East Sea, the Yalu and Tuman rivers, Mt. Baektu and its surrounding high-elevation areas, and the demilitorized zone (DMZ) function as the moat that protects the stronghold of North Korea. In addition, North Korea restricts the traffic of its own people and foreigners as well as commercial exchange and the transfer of information. North Korea and China have sixteen border entrances along the Yalu and Tuman rivers. South and North Korea are connected through Panmunjom, the West Corridor linking Kaesung and Paju, and the East Corridor linking Mt. Kumgang and Kosung. The problem here is that, in general, isolation is historically associated with underdevelopment (Modelski 1987; Zakaria 2009). North Korea's seclusion contributes to its political, economic, and cultural backwardness.

The governance centered on *Suryong* has resulted in state failure. The policy of prioritizing regime security has contributed to a distorted distribution of material resources. Without any criticism from below in the decision-making processes, the North Korean leadership has continued to make ill-fated decisions that have been detrimental to the welfare of the state and of society in general. Under this type of governance, the people have been forced to be mere objects of exploitation and mobilization. A cure for the problem of such a fortress state can only be found in redressing the imbalance of power between the state and society.

The Stalin Regime, the *Yuil* Regime, and the *Songun* Regime

The North Korean fortress state has for the past sixty years maintained its basic governance through its monopoly of power and planned economy and yet has evolved through three different regimes; that is, the Stalin regime, the *Yuil* (the one and only) regime, and the *Songun* (military-first) regime. Figure

Table 1 Regime Changes in North Korea as a Fortress State

	Stalin regime	*Yuil* regime	*Songun* regime
Major events	February 1948: The KPA established September 1948: The DPRK established 1950–53: The Korean War August 1956: The August Anti-Party Incident July 1961: Treaties of Cooperation and Mutual Aid with the Soviet Union and China December 1962: The KWP Central Committee adopts the Economy-Military Co-advance Line	January 1969: The military hard-liners are purged December 1972: Constitutional revision February 1974: The KWP Central Committee appoints Kim Jong-il as an official successor 1989–91: The Cold War ends July 1994: Kim Il-sung dies 1995–97: The period of the Arduous March	October 1997: Kim Jong-il appointed as General Secretary of the KWP September 1998: Constitutional revision; Kim Jong-il renamed Chairman of the NDC June 2000: Kim Dae-jung–Kim Jong-il Summit July 2002: The July 1 Economic Restructuring Measures October 2006: The first nuclear test October 2007: Roh Moo-hyun–Kim Jong-il Summit May 2009: The second nuclear test
Regime orientation: revisionist vs. status-quo oriented	1948–53: Revisionist; 1954–60: Status quo-oriented; 1961–68: Revisionist	1969–96: Status-quo oriented	1997–Present: Revisionist
Major characteristics	A typical Soviet regime is born Collectivization of the means of production	The *Yuil* system is established The regime erodes due to inefficiencies of isolation and planning The world socialist system collapses	*Songun* Politics emerges Efforts toward nuclear armament Coexistence of markets and planned economy

1 shows the basic characteristics of the three regimes.

Following the defeat of Japanese imperialism, the Korean People's Army (KPA) was established in February 1948. The DPRK was born in September of that year, and was later nourished into a typical Soviet regime. During this period, North Korea struggled to build a model socialist state under the guidance of the Korean Workers' Party (KWP). A number of factions—the Manchurian faction, the Soviet faction, the Yenan faction, and the Domestic faction—vied for total control of state power. Between the establishment of the DPRK and the end of the Korean War, North Korea pursued a revisionist path, calling for a unified state under the red flag by force. Following the end of the Korean War, North Korea turned into a statusquo-oriented power, focusing on the postwar rehabilitation of the national economy. In the 1960s, North Korea again turned revisionist as it resumed an interest in revolutionizing the whole peninsula by force.

As North Korea was building a *Yuil* regime, its revisionist tendencies went dormant from the late 1960s. The Manchurian faction seized total control of state power, effectively eliminating potential challenges to power from the outside, and the personal cult of Kim Il-sung rose to new heights in this period. Though masterful in tightly controlling its people, the *Yuil* regime proved to be quite unsuccessful in economic management. As it stuck to inefficient autarchy and central planning, its performance began failing and falling behind. With the backdrop of the collapse of the Cold War era, Pyongyang was hit hard by an economic breakdown and mass starvation known as the "Arduous March" (Noland 1997).

Since 1998, as Kim Jong-il rose to power, Pyongyang has again ventured to become a revisionist power. *Songun* North Korea, spearheaded by the National Defense Commission (NDC), seeks to compensate for its inferiority in conventional forces with the development of asymmetrical weaponry such as missiles and nuclear weapons. The *Songun* regime is distinguishable from the *Yuil* regime in terms of the nature of state-society relations. Under the *Yuil* regime, the dominance of the state over society was maintained. Under the *Songun* regime, however, as the state is no longer capable of providing basic means of survival for its people, state authority is seriously eroding. The comprehensive surveillance system is weakening. State failure in economic governance has resulted in the sprouting of markets nationwide. As women are more responsible for supporting their families, traditional patriarchy is fast losing ground (Lankov 2009).

What Is to Be Done?

Advancement comes from creating standards, and new standards are derived from innovation ahead of others. Innovation is the sum of invention and introduction (Nordfos 2009, 4). In order to achieve the status of an advanced nation, many requirements need to be met on various levels. First of all, at the individual level, community members who are capable of critical, rational, and creative thinking are needed (Dyer, Gregersen, and Christensen 2009).

At the societal level, there needs to be recognition of private property rights, fair competition among and multiple opportunities according to the merits of members, the rule of law, and good institutions of governance. These characteristics in coordination will create an environment where healthy and imaginative members are nurtured. Basically, advanced societies are open societies that freely import external information and human and material resources and export internal ideational and material resources outside. Freedom of transportation, communication, and commerce is the sine qua non of advancement. There needs to be a balance of power in state-society relations. The state needs to exercise leadership that facilitates the dynamism of society, and society needs to possess power that allows its freedom from the state and at the same time checks the state. At the international level, competition and active trade with other states are also positive forces for social and economic development (Zakaria 2009, chap. 3).

When these conditions are met at the individual, state, societal, and international levels, there arises a dynamism that becomes the basis for multiple innovations. Creative individuals, state leadership, and internal and external competition become the foundation for innovation in social management and technology, and innovations function as the basic endowment for advancement. We can sum these factors as:

creative individual + fair society + effective government + international competition
→ innovation (technological breakthrough)
→ advancement

North Korea today shows characteristics that are far from the general conditions of this formula for advancement. First of all, it seems that by and large the North Korean people, who have been accustomed to the teachings of state ideology, do not display critical and creative thinking and, rather, show reactive thinking in order to survive. The closed nature of North Korean society severely restricts its people's opportunities for creative thinking through comparison and analogy. The movement of information, material goods, and the people is highly constrained. There is little room for fair competition among the members for upward social mobility.

The end point of all the North Korean deficits is the state itself. The monopoly of power as manifested in the *Suryong* system lies at the heart of all the problems. No political order is free from the classic governance dilemma of "who guards the guardians?" as raised by Plato (Keohane 2001). The North Korean case suffers from the issue of "who guards the *Suryong*?" as he is responsible for every aspect of people's concerns. The North Korean governance system is devoid of mechanisms of checks and balances when the *Suryong* makes a mistake. North Korea seeks to face this problem with the belief that the *Suryong* never errs. Unfortunately, however, the *Suryong* does err, and the personalization and monopolization of power have created immense problems in North Korea's modern history. Therefore, a cure for North Korea's ills will be found in the transformation of malfunctioning mechanisms of governance and the installment of the normal operation of party apparatus (including the Politburo and the Central Committee) and of the cabinet, institutionalization of the power succession, and the separation of the party and the state from the Kim family. In other words, normalizing the exercise of power in Pyongyang means finding a way to balance and check the *Suryong*'s absolute power.

The second dilemma that North Korea faces is that the fate of the fortress state and the advance of reform efforts are in a reverse relationship; that is, as the pressure for reform deepens, the sustenance of the fortress state becomes increasingly more formidable, and therefore the state is quite unwilling to succumb to the pressure of opening and reforms. These two dilemmas need to be solved in a way that allows the *Suryong* and the fortress state gradually to fade away as reform and opening measures are gradually introduced. Just as some aspects of the state in North Korea are over-present, there are areas where the state is wanting. Where there needs to be state care, it simply does not exist, which is a form of state failure. As it stands now, the North Korean fortress state is unable to solve the people's need for access to food as well as for education, health, and welfare that are needed for basic subsistence. It is urgent that the state resume its normal capacities in these areas. The North needs to produce multiple institutions that will ensure the normalization of state capacities and sustainable governance for its people. The final aim of those institutions should be the guarantee of security, liberty, welfare, and justice for the members of the community. The procedures for the operation of governance should be based on accountability, participation, and persuasion, not unaccountability, exclusion, and propaganda and coercion as manifested in the current North Korean style of governance (Keohane 2001).

In the end, there is no bright future for the fortress state, and, therefore, North Korea needs to escape its current status and proceed toward the amphibious state via the bridge state:

Fortress State → Bridge State → Amphibious State

The bridge state refers to one that connects one state to others or one region to others that possess different characteristics. More specifically, the North Korean bridge state will connect continental powers and maritime powers through railroads, highways, and energy pipelines and assume the role of boosting reconciliation and cooperation among the regional members. The bridge state still bears the strong traits of a land power and yet is on the road toward a gradual transformation of shedding its authoritarian and closed characteristics and developing the liberal and open identities of a sea power. The concept of the bridge state was widely used under the Roh Moo-hyun administration to refer to the characteristics of the ROK. President Roh, under the slogan of the Northeast Asian Era Initiative, sought to build the Korean Peninsula Peace System and the Economic Community and the Security Community in Northeast Asia. He thought that South Korea needed to play the role of a bridge state connecting continental powers and sea powers in the fields of security, economy, and culture. I use the term "bridge state" to refer to North Korea, and the contention is that Pyongyang needs to turn into a bridge state before Seoul plays the role of a link (Park et al. 2006).

At some point in the future, North Korea will transform from a bridge state into an amphibious state. The amphibious state possesses the strengths of both a land power and a sea power. North Korea needs to shed its exclusively continental orientation and acquire the merits of a sea power, which include an explorative maritime orientation, interdependence, and an externally oriented economic development strategy. In order to solve its chronic problems, North Korea needs to govern in a system without a *Suryong*. The fortress state under his leadership has sought omnipresence, omniscience, and omnipotence. During the *Yuil* regime, Pyongyang had been somewhat successful in pursuing omnipresence, whereas the goals of omniscience and omnipotence had not been met. In the *Songun* regime, even omnipresence itself is increasingly eroding.

In the future, North Korea needs to curtail its overblown activities in politics and the military and to put more energy into the issues of the economy, the environment, culture, human rights, and the information and knowledge sectors. North Korea needs a balanced strategy that seeks to strengthen both its hard power and its soft power. Such advancement will be possible when it has become attractive enough as a homeland to prevent the exodus of its own members and to invite more creative minds from abroad so that its own soft power will function as a glue holding members of the society together. North Korea needs to overcome the distorted distribution of resources under the mandate of military-first politics and instead practice efficient distribution of resources under a new motto of economy first.

North Korea's Road to Advancement

Kim Jong-il is tenaciously standing by his military-first politics. It is highly unlikely that military-first rhetoric and practice will be discarded while he is in office. Therefore, I start with the premise that as Kim fades away, a comprehensive restructuring of North Korea and its society will become feasible. The new leadership that replaces Kim Jong-il will be able to launch new programs of state management from a fresh angle. North Korea will then be able to undergo a process of decay, transition, and transformation. The following assumptions undergird a step-by-step revolution of the North Korean system:

- The probability of an implosion or explosion of the North Korean system is not unusually high despite its contradictions, frailty, undesirability, and backwardness.
- The sudden collapse of North Korea is not desirable, because such a development would most likely cause unbearable and excessive pain not only for the North Korean people but also for their neighbors.
- The maintenance of North Korea as it now stands will become increasingly difficult due to the internal and external pressures on its system.
- The transformation of North Korea will not follow a linear trajectory. It will reflect a learning process of trial and error. North Korea will muddle forward through progress, backpedaling, and stasis.
- The primary agents of the North Korean transformation will be both North Korean elites and the North Korean people.
- North Korean changes will be accelerated and facilitated as they are met by a favorable international environment. As such, they will assume the characteristics of coevolution from both inside (unit) and outside (system) (Porter 2006).
- The momentum for North Korea's evolution will grow as the DPRK successfully accomplishes its initial tasks and copes with the risks associated with reforms.
- The North Korean evolution will take place in parallel in the sectors of politics, the economy, the military, society, culture, human rights, and

science and technology.

- The rise of an effective government and proactive society will magnify the scope of the changes.
- As time goes by and the experience of evolution accumulates, North Korea will be able to approach the goal of achieving an advanced stage.

I believe that the incremental transformation of the North Korean fortress state is both desirable and feasible. In what follows, I will introduce the basic characteristics of each stage along the path of advancement.

The Decline of the *Songun* Regime

As long as Kim Jong-il stays in power, North Korea will follow the *Songun* line. During this period, the *Suryong* governance as it is currently practiced is likely to be maintained, and hence the contradictions of the *Songun* system will mount and the people will keep suffering. The dictator, agitated by both his achievement deficit and his legitimacy deficit, will keep making enemies from within and outside and maintaining his desire to control and mobilize the people on his behalf.

Following the breakdown of socialism in the Soviet Union and the Eastern European countries, North Korea has been struggling to revive its ailing economy but is unwilling to engage in full-scale reform and opening. Between 2000 and 2004, it adopted some reform-oriented measures. In 2002, North Korea promulgated the so-called July 1 Economic Restructuring Measures and, in 2003, allowed opening of general markets nationwide. Under Premier Park Bong-ju, appointed in September 2003, reformist efforts were consolidated as markets and planning coexisted. Starting from early 2005, however, conservatives, unhappy with a series of liberal policy measures, began expressing their reservations and discomfort and fought to contain and roll back the spread of markets and reimpose state control (Park 2009).

For a long time, the state has been unable to provide the fruits it has promised its people. As the gaps widen between the lofty goals of *Juche* and *Songun* on the one hand and dire realities on the other, room for potential conflict between the rulers who promote military first and the people who desire economy first is growing. As people's basic needs for survival are not met, the gaps between the rich and the poor are increasing, and as people acquire additional information about life outside the walls of the fortress state, the legitimacy of the North Korean regime will continue to fall. The state and the people are drifting apart as they respectively pursue different goals.

Who will succeed Kim Jong-il is perhaps the most important issue for the first stage and will greatly affect the future of not only North Korea but also the Northeast Asian regional order. When a leader himself has created the ruling party and the process of power transfer is not fully institutionalized, the elites often rely on hereditary succession, which reduces uncertainties and prevents the development of a power vacuum. Between 1946 and 2006, nine cases of hereditary succession have been reported within autocracies. In the case when a son inherits power from his father, both the powerholder and his designate are motivated by a desire for safety. The supportive elites tend to favor a smooth transfer of power over an unpredictable power struggle. Hereditary succession can be seen as more attractive than a regime collapse or a power vacuum. It can be interpreted as a rational choice that brings benefits to both the successor and the elites (Brownlee 2007).

It has been reported that Kim Jong-il handed down an instruction naming Kim Jong-un as his heir on January 8, 2009, to the KWP's Department of Organization and Guidance. The KWP's Department of Organization and Guidance and Department of Propaganda and Agitation, the KPA's General Political Department, and the NDC are spearheading the succession process. It seems that the North Korean leadership has come to the conclusion that the next leader should come from the Baekdu bloodline to ensure the survival of North Korea's unique *Yuil*-style socialism (Koh 2008; Cheong 2009). The Meeting of Workers' Party Convention in late September 2010, which was the first to be held in decades, became a coming-out party for heir apparent Kim Jong-un. He is currently a four-star general of the Korean People's Army and vice chairman of the Party's Central Military Commission (*JoongAng*

Daily 2010). The success of this attempt at hereditary succession may hinge on the unity of the elites of Pyongyang and the direction of the North's nuclear standoff with the outside world.

During this period, North Korea will be unable to find a solution for its system deficiencies and will stick to the conservative policy of ensuring regime survival under the military-first slogan. Externally, a tug of war between North Korea and other powers involving denuclearization and economic assistance will continue. The speed and depth of denuclearization will seriously affect the process of North Korea's efforts for advancement. The resolution of the nuclear issue should be dealt with in stages 1and 2. If the nuclear issue is resolved in these stages, then in stage 3, the North Korean leadership will be relatively free to pursue the goal of economy first and the people first. The faster Pyongyang frees itself of its nuclear shackles, the freer it will become in pursuing reform and opening.

The Coming of the "Enlightened *Suryong*"

Stage 2 will be a transitional period, in which an "enlightened *Suryong*" will emerge. As Chairman Kim Jong-il "retires," a new leadership will strive to find solutions to the problems of the military-first period and to present a grand strategy for the nation's future. In order to uphold the continuity of the North Korean–style revolution, it is unlikely that Pyongyang will completely undo its past revolutionary traditions. But at the same time, the new leadership will strive to introduce more pragmatic programs that are distinguishable from those of the past leadership to win the hearts of the people and to earn legitimacy. The regime in this period will be a hybrid that shares the characteristics of both military-first policies and reform. A leadership change that replaces Kim Jong-il will be accompanied by regime change as well as generational change in the elite membership. The enlightened *Suryong*, though he is part of the bloodline of the Baekdu family, is likely to be the "last *Suryong*," who no longer passes on power to a descendant.

The new leadership will embark on relatively mild and limited reform and opening that is not as drastic as that of the next stage but notably different from the previous stage. A new *Suryong* will be unable to wield the absolute power of his predecessors due to his youth and lack of experience, as well as the changed international environment. He will be more interested in propagating new political thinking that includes experiments for economic growth. The enlightened *Suryong*, perched between the conservatives and the reformists, will play the role of a balancer who coordinates the interests and policies of both sides. This means that the nature of *Suryong* governance will be changed drastically, now that the former *Suryong* who presided over the people but was never responsible for anything no longer exists.

During this stage, the generational change and the diversification of the core elites will take place. A new generation of elites, recruited from various quarters of society, will replace the old ones. Technocrats will overwhelm old-hat ideologues. As the functions of party apparatus normalize, the focus of power will increasingly move from the *Suryong* to the party. The Supreme People's Assembly (SPA), a legislative organ of the DPRK, will increase its power of supervision over state apparatuses and the cabinet will be able to exert power more independently from the KWP.

The primary task for stage 2 is to find a formula of economic growth. The state will shift its orientation from regime survival to economic development. North Korea will embark on a reform and opening on a par with that of China and Vietnam. It will try to achieve a balance between markets and planning and prioritize light industry, tourism, and information technology.

The nature of North Korea will change from a fortress state to a bridge state. Pyongyang will play the role of a connector linking the South's maritime states and the North's continental states through energy pipelines, transportation, and communication. North Korea will open wide its doors to international society. In addition to Najin, Sinuiju, Kaesung, and Kumgangsan, North Korea is likely to open Wonsan, Heungnam, Haeju, and Nampo to international investment. As North Korea expands international exchanges, it needs to abide by global standards and stop engaging in illicit activities. To compensate for the income lost from unlawful businesses, Pyongyang will lean on passage rent in the form of transit fees over the railroads, highways, and energy pipelines that run through North Korea's territory.

State-society relations will be transformed in stage 2. The state will endorse private economic activities, including family farming, cottage industry, and other commercial activities and individuals' accumulation of wealth, though the collective ownership of the means of production will remain the norm. Accordingly, state intervention in people's private activities will decrease relatively. The effectiveness of state propaganda will also decrease as people are exposed to more outside information. The state needs to resort to material incentives to influence its people, because its official ideology cannot attract people's minds. Therefore, the success of this period will depend on the quality and quantity of material resources the state can produce for its population.

The North Korean leadership in this stage needs to be mindful of a number of daunting challenges. There exists the possibility that the elites may split into two quarreling groups of reformists and conservatives. As noted previously, the enlightened *Suryong*'s primary role is to seek a balance between the two. He needs to ensure that the policy debates do not flame into bloody power struggles. A split between the civilian elite and the military elite is another possibility. The military elite may be divided between the central military of the National Defense Commission and the Ministry of People's Armed Forces and the field armies of the countryside. The enlightened *Suryong* needs to be a compass guiding the reform efforts and at the same time block the rise of resistance from the enemies of reform.

The drifting apart of the regime and the masses may intensify in stage 2. The people will be armed with more information as society opens up and people's demands will increase as well. If social safety nets such as pensions, health care, welfare, and education are not properly instituted and managed and the supply of the goods and services that will satisfy people's needs is insufficient, social stability may be in danger. The state and the masses will both be novices regarding the liberalization of society. The state has so far been indifferent to the demands from below. The masses are not accustomed to turning their discomfort into political action. When individuals are equipped with the Internet, mobile phones, and copying machines, the state will find it more difficult to control them effectively. As the nature of state-society relations changes, the need will arise for the introduction of new institutional knots that tie the state and its people together.

When initial economic growth hits a wall and starts reversing, and the income gap between the poor and the rich widens, one cannot rule out the eruption of violence among the masses as a sense of relative deprivation escalates (Gurr 1970). As control over the people loosens, the chances for "voice," or resistance, and "exit" will rise (Hirschman 1970). In stage 1, exit was the only option as voice was tightly suppressed by the state surveillance machinery. In stage 2, the possibility exists for voice and exit to mount simultaneously. To prevent this from happening, the North Korean leadership should produce a number of qualified programs to appeal to the loyalty of the people. The success of the enlightened *Suryong* will depend on whether he is able to control the eruption of diverse demands during the transitional period and to manage the conflict of interests among different quarters of society.

For stage 2 to become a stepping-stone linking stage 1 and stage 3, it is imperative that the nuclear problem be resolved once and for all during this period. As Pyongyang declares its resolution to undo its nuclear weapons programs at the end of stage 1, stage 2 will entail the process of executing actual denuclearization. North Korea's denuclearization is the sine qua non of its advance to developed-nation status. The faster Pyongyang gets rid of its nuclear ambitions, the faster it will be able to truly advance. North Korea's denuclearization will require the joint operation of bilateral bargaining between South Korea and North Korea and between the United States and North Korea and the multilateral efforts of the Six-Party Talks. North Korea's denuclearization will proceed along with changes in the regime type of North Korea and in the nature of the international environment. As denuclearization proceeds, North Korea's relations with neighboring powers will begin to improve and efforts for building a Korean Peninsula peace system will take off.

Transformation into Democracy

On the basis of the trial-and-error experience of the second stage of transition, North Korea will be able to launch a full-scale transformation in the third stage. In stage 3, North Korea will undergo a systemic transformation, and North Korean–style democracy will be installed. North Korea's past, elite choices, and external influences will all have an imprint on its democratic style.

In the early phase of stage 3, North Korea's unique *Suryong* governance will finally come to an end as the next leader will no longer be selected from the so-called Baekdu bloodline base. The rule of law will be established, effectively replacing rule by the *Suryong*, and the Politburo and the Central Committee will take center stage as democratization of party politics sets in. Even though the monopoly of power by the KWP is sustained, the spread of democratization within the party will make room for more transparent decision making. Collective leadership will replace the *Suryong*, and power succession will be institutionalized as a new leader is selected by elite consensus. The institutionalization of the power succession will have the effect of decreasing unnecessary political transaction costs in times of leadership change.

During this period, reformists will outnumber conservatives among the elite, and the reform-oriented leadership will proactively pursue reform and opening on the principle of economy first. During this period, North Koreans will enjoy a level of economic freedom similar to that in Vietnam and China, and with time, liberty will expand and liberalism will prevail. The balance between markets and planning will incrementally change in favor of the markets. Private farming will expand at the expense of collective farming and state-owned farming, both of which will finally be abolished in due course in stage 3. The numbers of state-owned enterprises will diminish.

In the second phase of stage 3, one-party rule by the KWP will come to an end. Multiple political parties besides the KWP will coexist and compete with one another to become the ruling party. These parties will develop different worldviews and programs to persuade the hearts of the North Korean people. Each party will present different programs regarding its relationship to the traditions of North Korean socialism and the outside world. Parties will also present different ideas about how society is to be managed and how the economy is to be operated.

The main task of North Korea in stage 3 is to emerge from passive reactions to develop a problem-solving attitude and to create its future proactively and aggressively. In addition to being a compass for the nation's future, the state needs to create an environment where the dynamism of society is sustained. Society needs to possess veto power over the state's undue intervention. As the rule of law becomes firmly grounded, the legislative, the executive, and the judiciary branches will check and balance one another, and none will be able to dominate the others decisively. Lawmakers will be chosen in free and competitive elections. They will wield real lawmaking authority and will be able to oversee the activities of the state. The judiciary will conduct fair trials without political considerations or fear of political reprisal.

In order for North Korea to enter the ranks of the advanced states, individuals' property rights and the right of relocation and freedom of travel should be protected. Citizens need to have the freedom to choose their leaders without external interference and to be able to monitor their leaders' public activities. The guarantee of the basic rights of the people and the consolidation of procedural democracy should be firmly planted so that the people have ample motivation to work, their creativity is fully realized, and a just society is realized (Diamond 2008). The spread of the market economy combined with the consolidation of democracy will have an upward spiral effect in transforming the basic characteristics of the North Korean system.

During its race for transformation, North Korea will have to face multiple challenges from both inside and out. As time passes, external challenges will become more formidable. In the last phase of stage 3, the forces pushing North and South closer toward union will be difficult to ignore. In this stage, North Korea's primary sponsor as well as its latent challenger will be the ROK. To compete with the ROK, which will have long maintained an open society by that time, will be a daunting task for the DPRK, which will be barely beginning to embrace the international community. After all, South Korea has successfully achieved the goals of democracy and industrialization and is about to cross the line into advanced-nation status. North Korea is a reform tyro who has achieved little yet. It is urgent for Pyongyang to acquire sector-specific superiorities and create the centripetal force needed in order to tie the people to its side.

In stage 3, cooperation between the DPRK and the international community will blossom. As the thorny nuclear issue is resolved near the end of stage 2, diplomatic normalization between Pyongyang and Washington and between Pyongyang and Tokyo will be realized and assistance from international financial organizations and neighboring countries will burgeon. Foreign assistance will not be restricted to the Special Economic Zones but will be spread all over the North Korean countryside. North Korea, as it takes off its identity as a fortress state and transforms into an amphibious state through a bridge state, will become more and more intimately connected with the international community.

As North Korea strengthens its ties with South Korea as well as with other regional powers, the politics of the Korean Peninsula will assume more complex and dynamic features. In East Asia, competition will intensify as great powers will want to expand their zones of influence and small and middle powers will struggle to counter them. In this space, for the first time since the end of Japanese colonial rule, North and South Korean diplomatic collusion, in which the two Koreas cooperate with each other against other regional powers with an aim to advance the interests of the Korean nation, can emerge. Until now, Seoul and Pyongyang have been engaged in zero-sum diplomacy in which each has sought to cooperate with other regional powers at the expense of the other party. As the experience of South-North diplomatic coordination accumulates, the two Koreas will reach the stage of diplomatic union in which each maintains its separate political, economic, and social system but the duo pursues a unified diplomacy. The experiment of diplomatic collusion will function as a catalyst for inner social and political integration between the two Koreas and pave the way for a final unification. That is, following stage 2, the politics of the Korean Peninsula will be realized as follows: deepening of economic cooperation → diplomatic union → economic integration → social integration → political unity.

Conclusion

North Korea needs to escape from its isolation and rise to the status of an advanced state. What would it take for Pyongyang to turn into an advanced state through its own efforts? With military-first politics, North Korea will be unable to meet the challenges posed by the trends of the globalization plus information era.

I have proposed that North Korea's development as an advanced state will occur in three stages. In stage 1, North Korea will fail in making a transition toward democracy and advancement if it sticks to military-first policy. In stage 2, the advance of the enlightened *Suryong* will pave the way for partial reform and opening. Stage 3 is set for the demise of the *Yuil* system and the victory of the reformists over the conservatives. With the development of innovation-friendly individuals, society, institutions, and state, North Korea's competitiveness as a whole will rise and Pyongyang will draw nearer to the goal of advancement. ■

■ References

Brownlee, Jason. 2007. "Hereditary Succession in Modern Autocracies." *World Politics* 59, 4: 595–628.

Cheong, Sung-jang. 2009. "*Post Kim Jong-il Si-dae Bug-han Gwon-lyeog-che-gye-ui Byeon-hwa Jeon-mang*" [Prospect of Changes in North Korea's Power System in the Post-Kim Jong-il Era]. *Sejong Policy Studies* 5, 1.

Chesterman, Simon, Michael Ignatieff, and Ramesh Thakur, eds. 2005. *Making States Work: State Failure and the Crisis of Governance.* Tokyo: United Nations University Press.

Diamond, Larry. 2008. "The Democratic Rollback: The Resurgence of the Predatory State." *Foreign Affairs* 87, 2: 36–48.

Dyer, Jeffrey H., Hal B. Gregersen, and Clayton M. Christensen. 2009. "The Innovator's DNA." *Harvard Business Review* (December): 61–67.

Gurr, Ted Robert. 1970. *Why Men Rebel*. Princeton: Princeton University Press.

Hirschman, Albert O. 1970. *Exit, Voice, and Loyalty: Responses to Decline in Firms, Organizations, and States.* Cambridge, MA: Harvard University Press.

JoongAng Daily. October 12, 2010.

Keohane, Robert O. 2001. "Governance in a Partially Globalized World." *American Political Science Review* 95, 1: 1–13.

Koh, Yu-hwan. 2008. "*Kim Jong-il Geon-gangbyeon-su-wa Hu-gye-gu-chug-ui Scenario*" [The Scenario Concerning the Health of Kim Jong-il and the Succession Issue]. *KDI Review of the North Korean Economy* (September).

Lankov, Andrei. 2009. "Changes to North Korea's System of Social Control." *North Korea Newsletter* 77 (October 22).

Lee, Cho-won. 2008. "*Chwi-yag-gug-ga Model-gwa Bug-han-e Dae-han Gae-bal-hyeob-lyeog Yeon-gu*" [A Study on the Fragile State Model and Development Assistance for North Korea]. *North Korean Studies Review* 12, 2: 289–320.

Modelski, George. 1987. *Long Cycles in World Politics*. London: Macmillan Press.

Noland, Marcus. 1997. "Why North Korea Will Muddle Through." *Foreign Affairs* 76, 4: 105–118.

Nordfos, David. 2009. "What Is Innovation?" *eJournal USA* 14, 11: 4–7.

Park, Hyung-joong. 2009. "*Hwa-pye-gyo-hwan-jo-chi-ui Pa-janggwa Jeon-mang: Jeongchi-gyeongje-hag-jeog Bun-seog*" [The Impact and Prospect of Currency Reform:

Political Economic Analysis]. *KINU (Korea Institute for National Unification) Online Series* (December 10).

Park, Jong-chul, et al. 2006. *Han-gug-ui Dongbug-a-si-dae Gu-sang: I-lon-jeog Gi-cho-wa Che-gye* [Korea's Plan for the Era of Northeast Asia: Theoretical Foundation and System]. Seoul: Oruem.

Park, Sun-won. 2002. "*Kim Jong-il Si-dae Bug-han-ui Byeon-hwa: Jin-hwa-lon-jeog Jeob-geun*" [North Korean Reform in the Kim Jong-il Era from the Perspective of "Change within the System," 1997–2001]. *Korean Political Science Review* 36, 3: 155–173.

Porter, Terry B. 2006. "Coevolution as a Research Framework for Organizations and the Natural Environment." *Organization and Environment*.19, 4: 479–504.

Shin, Ji-ho. 2002. "*Bug-han-che-je-byeon-hwa 3 Dan-gye-lon: Dae-bug-jeongchaeg-e-ui Si-sa*" Three Steps of Change in the North Korean Regime: Implications for North Korea Policy]. *Korean Political Science Review* 36, 4: 103–119.

Vinci, Anthony. 2008. "Anarchy, Failed States, and Armed Groups: Reconsidering Conventional Analysis." *International Studies Quarterly* 52, 2: 295–314.

Woo, Seongji. 2008. "*Nambug-han Gwan-gye 60 Nyeon Bun-seog: 1948–2008*" [Analysis of Sixty Years of Inter-Korean Relations: 1948–2008]. *Review of International and Area Studies* 17, 2: 121–156.

Zakaria, Fareed. 2009. *The Post-American World*. New York: W. W. Norton.

CHAPTER 3

Coevolution Strategy for North Korea
Diplomacy

Chaesung Chun

Introduction

AFTER THE COLD WAR ENDED AND THE FORMER SOVIET BLOC BROKE down, North Korea's rival, South Korea, established diplomatic relations with the former Soviet Union and China, which put North Korea on a difficult journey for survival. Witnessing long-standing communist regimes that had worked alongside Pyongyang to construct socialism helplessly collapse or change over from communism under external pressure or pressure by the people, North Korea found itself in a desperately dangerous situation in which it had to sustain its socialist system and regime, and worry about not just a transition but an absorbed unification by the South.

With the fall of the socialist bloc, the U.S.-led liberal democratic bloc's development accelerated even further. In the post–Cold War environment, not only the new structure of power but also the development of the massive trends of globalization, information proliferation, and democratization confirmed socialism to be a system lacking the open, democratic, and creative qualities it needed to cope with these changes. In addition, liberal democracy developed in tandem with a market economy. At the same time, as the United States, who was the longtime adversary of North Korea, built an unrivaled hegemony with unprecedented qualities of leadership, North Korea felt ever more threatened.

The Kim Il-sung regime, which was searching for a survival strategy throughout 1991 and 1992, stirred up the first nuclear crisis in 1993, and the Kim Jong-il regime that followed in 1994 tried to escape its post–Cold War crisis by freezing its nuclear program but firmly promoting military-first politics. In the process, North Korea focused on aligning its internal system and strove to provide an opportunity for economic development, on the one hand, while concentrating on creating a diplomatic environment in which North Korea could survive on a long-term basis, on the other. On the basis of its bilateral diplomatic relationship with the United States that was arranged after the first North Korean nuclear crisis, North Korea tried to put an end to the so-called "United States' hostile policy toward North Korea" while it sought to strengthen its self-defense capacity based on military power that revolved around military-first politics. In regard to South Korea, Pyongyang, while keeping itself from an absorbed unification, utilized Seoul's engagement policy to secure practical interests for the sake of future regime development.

Also, by improving relations with neighboring states such as China, Russia, and Japan, North Korea attempted to maximize economic and diplomatic support from them. At the same time, it established diplomatic relations with non-Asian states, including those in Europe, and pursued a practical diplomatic doctrine that could have economic and diplomatic benefits, even if only partially.

However, facing a more conservative U.S. foreign policy and the antiterrorism campaign after the 9/11 terrorist attacks in 2001, North Korea found itself in a much more unfavorable diplomatic situation. Eventually, in October 2002, the outbreak of the second North Korean nuclear crisis signaled that North Korea was reverting back to the rigid nuclear military-first diplomacy centered on its nuclear program and nuclear weapons. Within the framework of the Six-Party Talks, sporadic attempts to discuss North Korean abandonment of its nuclear program with accompanying economic aid continued, but North Korea did not have the strategic option of abandoning its nuclear program. Its neighbor states' security guarantees failed to satisfy its insecurities, and thus the North Korean problem based on the nuclear weapons issue is still going in circles without much progress.

The North Korean nuclear program is not a nuclear weapons issue but a question of the survival of North Korea as a political entity. North Korean nuclear weapons diplomacy is also not about nuclear weapons but about its survival and the overall political system based on military-first priorities. As long as North Korea's nuclear military-first diplomacy and military-first political system interact reciprocally, the individual evolution of North Korean diplomacy is impossible. And if North Korea's politics, economy, military, society, and other areas do not evolve in tandem, neither will its diplomacy.

Furthermore, North Korea's evolution without outside support cannot fulfill the goals that are desired. Neighboring states' foreign policy toward Pyongyang must also evolve at the same time to support and strengthen North Korea's domestic forces that advocate such advancement. North Korea and its neighboring countries' coevolution will alleviate North Korea's political concerns and, in the long run, create diplomatic agreements on and action plans for North Korea's role in Northeast Asia.

This chapter predicts that if North Korea pursues nuclear military-first diplomacy in order to support its "strong and prosperous state" policy, it cannot survive as a normal and developed state in the international community. North Korea must make a strategic decision to abandon its nuclear program and revive as a truly developed state, meeting twenty-first-century global standards. In order to achieve this status, the evolution of both North Korea's political system and its diplomacy must work simultaneously. Moreover, North Korea's evolution should be supported by the evolution of neighboring states' North Korea policy. I first examine the development of North Korean diplomacy and analyze why North Korea's military-first diplomacy will inevitably fail. Then I introduce the three steps of diplomatic evolution for North Korea, and through a discussion of the coevolution process of it and its neighboring states, explore the possibility of whether its overall system can lead to a normal and developed future.

Components and Evaluation of Diplomatic Strategy in the Military-First Era

Components of Diplomatic Strategy

Diplomacy in the military-first era has the strategic goal of serving the strategic purpose of military-first politics. Military-first politics, which made the military the most important pillar of the regime while it promoted the militarization of the overall society, is a method to maintain and further strengthen the survival capacity of the North Korean socialist regime after the end of the Cold War. Therefore, diplomacy in the military-first era, or, in other words, military-first diplomacy, aims to guarantee the survival of North Korea's "Great Leader"–centered socialist regime and focuses on creating a diplomatic environment in which North Korea can eventually become a "strong and prosperous state." As the authority and power of the military have been reinforced, the status and influence of the military in the decision-making process of foreign policy have only increased. Pyongyang, in deciding on its method of enforcing foreign policy, has frequently resorted to military action. Its decision to develop nuclear weapons to use them as an important diplomatic tool can also be understood as one product of the military-first political system.

This argument about North Korea's military-first politics is closely related to its evaluation of the situation of the international community and the direction of future foreign policy. In other words, this discussion analyzes the environment of military-first politics while finding the "fundamental questions arising in today's world" from the international political environment. North Korea proclaims the goal of its foreign policy as "protecting and adhering to *Juche* [Self-reliant] socialism, the bastion of global socialism, from the allied forces of imperialism, and completing *Juche* achievements" (Kang 2002, 7–8). Additional explanations of this statement read "as long as imperialism exists on the face of the Earth and the schemes of aggression by imperialists continue, we cannot submit to such forces. The ultimate triumph of *Juche* achievements can only be accomplished by unrelenting fights. Also, the fundamental spirit of military-first ideology is to fully invoke military, political and economic power to fight to the end against the imperialists who try to crush *Juche* ideology. Through this process, we shall independently build a new peaceful society based on the dynamic spirit of military-first ideology" (Kang 2002, 15). Such statements show the direction of North Korea's diplomatic strategy.

Military-first diplomacy more specifically comprises (1) securing a survival environment through strengthening military deterrence and negotiations with the United States, (2) practical diplomacy for acquiring economic resources, and (3) diplomacy with South Korea in which North Korea, in the name of unification strategy, acquires the necessary diplomatic resources to secure economic aid and an environment in which it can survive. On this survival environment, North Korea asserts that "only a strong military response to the hostile imperialist forces, that only pursue wars of aggression, can lead to peace." It also states that "it is military-first politics that will provide the peaceful environment necessary for our nation's unification" (Kang 2002, 268). Moreover, on acquiring economic aid, it explains that "when our republic suffered an unprecedented food crisis and energy shortage that damaged our overall economy because of U.S. economic sanctions and their schemes using military pressure, military-first politics was the force that drove the military to be the primary actor that vitalized the overall economy." It also explains that there is a need to "make the military the main force and mightily construct overarching socialism" (Chun 2004, 32). On its strategy toward South Korea, it argues that "military-first politics is the strongest measure of unification our generation can use to realize our nation's long-cherished wish for unification." It further states that "military-first politics finds its meaning in that it hinders the United States' attempts at wars of aggression and thus provides the fundamental setting for a peaceful environment for our nation's unification." It asserts that "this is all because military-first politics is a political method that provides deterrence [in] preventing war on the Korean Peninsula."

Such military-first politics is also implied in the revised constitution of North Korea. In the Eighth Constitution revised on September 5, 1998, the expression "Marxism-Leninism" in Article 4 of the Seventh Revised

Constitution was deleted. The phrase "repel foreign powers on a national scale" of Article 5 was deleted as well. Moreover, in the Eighth Revised Constitution, Article 17, "autonomy, peace, amity" were set as the goals of foreign policy and at the same time, the diplomatic stance was based on "our own version of socialism and strengthening of international solidarity based on autonomy." The fundamental direction of North Korean diplomacy has thus shifted (1) from revolutionary diplomacy to practical diplomacy, (2) from diplomacy focused on "liberating the South" to a protectionist diplomacy to maintain the North Korean socialist regime, (3) from encampment diplomacy to nonalignment diplomacy in the short run and to open diplomacy in the long run, and (4) from a self-reliant diplomacy between China and the Soviet Union during the Cold War to inducement diplomacy due to Sino-Soviet détente.

Such fundamental diplomatic orientations of Pyongyang clearly differ from its diplomatic strategies during the Cold War period. The Korean Workers' Party code and the former socialist constitution stipulate that "the immediate purpose of [the] Korean Workers' Party is to achieve complete socialist victory in the Northern part of the Republic and [finish] national liberation and the historic task of the people's democracy revolution on a national scale. The ultimate goal is to penetrate *Juche* ideology across the entire society and [construct] a communist society" (Preamble of the code). For this goal, North Korea adopted a platform to strengthen the three main revolution capabilities and increase diplomatic relations within the socialist bloc and the Third World through strong ties to the Soviet Union. The one continuity is that amid the lack of evidence that North Korea has given up on communist unification, it has not given up the political intention to put the entire Korean Peninsula under its influence by measures other than war. As Hak Soon Paik has pointed out, "In the historical, psychological, and substantial space of the 'Korean Peninsula,' the essence of North Korea's strategy, to win the competition between the two Koreas and secure a monopolized authority over the entire peninsula to build a unified nation, has not changed" (Paik 2003, 150).

The core goal of military-first diplomacy was diplomatic relations with the United States. Immediately after the end of the Cold War, North Korea focused on negotiating with the United States, and between December 1988 and December 1992, it pushed forward adviser-level diplomatic contacts in Beijing twenty-eight times (Kim 2002, 149–151). However, as North Korea realized that the required factors for its survival were increasingly lacking, it decided on walking the path of military-first diplomacy and then launched the first nuclear crisis in March 1993. Afterward, by agreeing to freeze its nuclear program in the 1994 Geneva Agreed Framework between Washington and Pyongyang, North Korea tried to make diplomatic room for its survival for the time being. After going through the first nuclear crisis and signing the Agreed Framework, the bilateral relationship between the United States and North Korea began, and until the DPRK-U.S. Joint Communiqué under the Clinton administration in 2000, desperate efforts to improve U.S.-DPRK relations, known as the "Geneva Agreed Framework system," continued for eight years. This relationship then experienced a year-long lull after the Bush administration came into office. In 2002, with the outbreak of the second nuclear crisis, this bilateral relationship entered the era of the Six-Party Talks, but without an effective breakthrough, it stagnated.

North Korea, based on its nuclear program, proceeded with its military-first diplomacy while trying to sign a peace treaty with the United States and to create the military and diplomatic environment necessary for its survival. After the 1994 Geneva Agreed Framework, U.S.-DPRK relations showed partial progress. On January 12, 1995, the U.S. State Department announced that it would ease the sanctions on North Korea, allow business and travel between the United States and the DPRK, grant permission to North Koreans to use U.S. finance institutions, and engage in the importation of magnesite from North Korea. During this process, North Korea continually emphasized the signing of a U.S.-DPRK peace treaty. While demanding the conclusion of a peace treaty, North Korea took measures such as expelling the delegation of the Czech Republic in 1993, establishing Panmunjom representatives of the Korean People's Army in 1994, and pulling out North Korean representatives from the Panmunjom Military Armistice Commission, to jeopardize the armistice system. The Four-Party Talks were held in March 1997, but after six talks ending in June 1998, the Four-Party Talks ceased due to North Korea's refusal to participate. This was the result of North Korea's repeated argument that the U.S.-DPRK peace treaty must be signed first.

North Korea promoted a diplomatic policy based on its nuclear program to fortify the requirements for its survival and to develop a "strong and prosperous state" by signing a peace treaty with the United States. The North asserted that securing the peace of the Korean Peninsula was important because, first, a treaty would "make the peace of the Korean Peninsula serve our nation's unification, second, [realize] peace through dialogue and negotiation, third, completely [withdraw] the U.S. troops from the Korean Peninsula, and fourth, strictly [denounce] the increase in military power and defense expenditure of South Korea" (Eom and Yun 2006, 120). Then North Korea further elaborates clearly on the peace treaty. "The military contest between the Republic and the United States ... end the hostile relations and based on reciprocal trust, conduct talks and contacts, and end the disreputable past ... the first step is no other than a DPRK-U.S. Nonaggression Pact." For these reasons, it argues that "first, as long as the U.S. military threat is not eliminated, North-South Korean relations cannot normally improve ... a Nonaggression Pact must be signed before the distrust and hostility between the DPRK and [the] United States can alleviate and guarantee each other's security ... talks must take place ... [would be] practical and meaningful." "Second, the signing of this pact will be [a] turning point in ultimately solving the problem of the Korean Peninsula. It will nullify the justification for stationing U.S. troops on the Korean Peninsula and will play the decisive role of opening the new phase of our nation's independent and peaceful unification by providing the objective guidelines for the withdrawal of U.S. troops" (Eom and Yun 2006, 227–233).

Such efforts reached a peak in 2000. In October 2000, Jo Myung Rok, North Korea's first vice chairman of the National Defense Commission, visited the United States and agreed to ending the hostile relationship, and Secretary of State Madeleine Albright visited Pyongyang and discussed improving U.S.-DPRK relations. In addition, the United States made basic agreements for the signing of the peace treaty and, in return, it clearly stated in the agreement the banning of launching missiles. The Joint Communiqué states, "Through historical highest-level talks, the DPRK and the United States agree that the circumstances of the Korean Peninsula have changed and for the strengthening of the peace and security of the Asia-Pacific region, the two states decided to take measures to fundamentally improve the bilateral relationship. With this understanding, both parties will relieve the tensions on the Korean Peninsula, change the 1953 armistice to a stable peace-guarantee system and both acknowledge that there are many ways, such as the Four Party Talks, to officially end the Korean War" (Korean Central News Agency 2000). Such discussions continued into the second Bush administration. "To resolve the rigid situation of the Korean Peninsula, we [North Korea] believe that signing a DPRK-U.S. Nonaggression Pact is a reasonable and realistic method for solving the nuclear problem. Through the Pact, if the United States legally promises nonaggression, including the use of nuclear weapons, we also have intentions to alleviate U.S. security concerns. For a small country such as ours, the criterion for all problem-solving is the elimination of threats to our sovereignty and survival." Such statements show that diplomatic relations with the United States are at the core of North Korean military-first diplomacy (Statement by the Spokesperson of North Korea's Ministry of Foreign Affairs 2002).

Since the launching of the Bush administration and the strengthening of U.S. nonproliferation and counterterrorism policies after 9/11, DPRK-U.S. relations have been going steadily downhill. North Korea describes the international situation of the 2000s, particularly relations with the United States, as follows: "To secure peace on the Korean Peninsula and to achieve an independent unification, we must oppose invasion, war schemes and eliminate threats of war. Under the justification of counterterrorism, U.S. and South Korean aggressive reactionaries are arousing tensions on the Korean Peninsula with anti-Republic and anti-unification schemes. In the international circumstances of increasing threats of a new war, we cannot think of our nation's peace and unification" (*Rodong Sinmun* 2002). Also, "after the launching of the Bush administration, the United States has pegged us as the 'Axis of Evil' and has declared it national policy to reject our system, and by designating us as a target of preliminary nuclear attack, it has publicly declared a nuclear war on us. The United States was already deliberately violating the Agreed Framework and in the meanwhile, it has accused us of new 'suspected nuclear development' and suspended heavy fuel oil aid, thus undisputedly violat[ing] the agreement. In response to our sincere proposals and serious negotiation efforts, the United States has responded with threats of 'blockades' and 'military response,' and has shown an arrogant attitude by insisting [on] 'talks but no negotiations'" (Government Statement of Democratic People's Republic of Korea 2003).

As a result, North Korea reverted back to its military-first diplomacy, based on nuclear weapons, in dealing with the United States, and after the second nuclear crisis broke out in 2002, U.S.-DPRK relations faced a stalemate. North Korea's nuclear military-first diplomacy has in fact strengthened, and as it has conducted its first and second nuclear testing, its diplomatic isolation and economic sanctions have been intensified.

If diplomacy with the United States is one axis of military-first diplomacy for securing survival, the other axis is practical diplomacy that seeks to maintain the economy and find sources of development. North Korea's attempts at practical diplomacy can be seen in its one-directional diplomatic relations. DPRK-Chinese relations have temporarily cooled after the establishment of diplomatic relations between South Korea and China. However, continuous exchanges have taken place, including the signing of the DPRK-China Economic Technology Cooperation Agreement in 1995, Chairman of the Presidium of the Supreme People's Assembly Kim Yong-nam's visit to China in June 1999, and the Foreign Minister of the People's Republic of China Tang Jia Xuan's visit to North Korea in October 1999. Particularly, North Korea normalized relations with China after Kim Yong-nam's visit to China in June 1999, and starting with Kim Jong-il's unofficial visit to China in May 2000, Kim Jong-il visited Shanghai in January 2001, and Premier Jiang Zemin visited North Korea. Through such summit diplomacy, North Korea strove to improve the relations between the two countries.

In tandem, Russo-DPRK relations also developed. As the Russo-DPRK Treaty of Friendship expired in 1995, North Korea, in the process of negotiating the new Russo-DPRK Treaty of 1997, deleted the automatic military intervention clause and support for the Korea Federalist system clause. After the initial signing in March 1999, North Korea signed the Russo-DPRK Treaty of Friendship, Good-Neighbor, and Cooperation Agreement. Afterward, the two countries entered a new phase in their relationship, and as seen in Vladimir Putin's visit to North Korea in July 2000 and Kim Jong-il's visit to Moscow in July–August 2001, they increased their efforts to strengthen their cooperation.

DPRK-Japanese relations have also changed according to DPRK-U.S. relations and general changes in circumstances. Both North Korea and Japan have seen fluctuations in their relationship in response to issues of history, reparations as an economic problem, nuclear weapons and missiles, ongoing problems of abductions and food aid, and *zainichi* (residing in Japan) Koreans. In its diplomatic policy toward Japan, North Korea has said that it is hard to improve relations because of the perception that Japan follows the United States, that Japan discriminates between North and South Korea, and emphasizes the problem of discrimination against the *zainichi* Koreans, the perception that North Korea needs Japan's capital and technology for North Korea to have an economic breakthrough, that North Korea can use DPRK-Japan negotiations to improve DPRK-U.S. relations, and that it can earn loans somewhat like compensations through establishing diplomatic relations with Japan.

From January 1991 to November 1992, North Korea and Japan held talks to discuss normalization of their relationship eight times, and after the Geneva Agreed Framework, the labor party of North Korea and the coalition government of Japan had agreed in March 1995 to resume talks for normalization. However, as North Korea tested its missiles in August 1998, its relations with Japan rapidly cooled, and in August 1999, North Korea asserted improvement in its relations with Japan and called for reparations for historical grievances. After the DPRK-U.S. Berlin Agreement in September 1999, the two countries resumed normalization talks on December 3, 1999, and on December 19, the two countries came to an agreement on issues such as the third visit to Japan, through Red Cross talks, of Japanese women living in North Korea.

North Korea also sought improved relations with South East Asian countries. In July 2000, it joined the ASEAN Regional Forum, reinstated relations with Australia (May 2000), and established diplomatic relations with the Philippines (July 2000), New Zealand (March 2001), and so on.

Improvement in relations with the European Union (EU) is a typical example that shows North Korea's practical diplomacy. After the mid-1990s, the relationship between North Korea and the EU expanded, and the EU supported North Korea with food aid, technical assistance, and medical support for flood relief in 1995. Also, the EU paid 75 million euros to the Korean Peninsula Energy Development Organization (KEDO) business by the end of 2000. People-to-people exchanges were also expanded as an EU delegation visited the Rajin-Sunbong economic trade zone in May 1998 and as a European Parliament delegation visited Pyongyang in December. After North Korea started the first political talks in December 1998, with the establishment of diplomatic relations with Italy in January 2000, it did the same with most of the European countries. With countries such as the United Kingdom, Netherlands, Belgium, Spain, Germany, Luxembourg, and Greece, it built normal diplomatic relations (Korean Association of North Korean Studies 2006).

The aspect that shows the results of North Korea's practical diplomacy is the amount of aid it has received. From 1995 to 1999, the total amount of supplies that South Korea and the international community provided to North Korea at no cost was 1,483,390,000 dollars, which averaged 300 million dollars a year. Just looking at the year 1999, the total amount that North Korea received as aid was 624,990,000 dollars, which equaled 5 percent of North Korea's 1998 GDP of 12.6 billion dollars, 43 percent of total volume of trade at 1.45 billion, and 112 percent of export volume at 560 million dollars (Kim 2002, 149–151).

Evaluation of the Diplomatic Strategy

The post–Cold War diplomacy of North Korea can be characterized as a combination of realism and principle-oriented justification, which can be assessed under the framework of domestic politics and foreign policy. While its external relations exhibit both moralistic and idealist notions of socialism, North Korea's arguments against imperialism are closer to a moral rejection based on the ideals of self-rule, peace, dignity, and survival. Since such a framework is intertwined with the ideological basis buttressing internal legitimacy, it can be understood as a conjoining of both external justification and internal realism as well. This is most visible in the regime's policy toward the United States, where continued vigilance and rejection against the United States have been the underlying principled approach.

However, there are also signs of transformation in North Korea's external policy that can be understood as a change toward a more realist approach from an ideological orientation. For instance, despite how the regime had been criticizing U.S. policy toward North Korea, remarks from the October 2000 Joint Communiqué that expressed its preference for "negotiation over deterrence," recognition of improving ties as a "natural goal," reaffirmation of relations with the United States to be "based on the principles of respect for each other's sovereignty and non-interference in each other's internal affairs," and noting "the value of regular diplomatic contacts, bilaterally and in broader fora," seem to suggest that new change is inevitable. In that sense, North Korean diplomacy can be assessed as a realist alignment by the regime, seeking to adapt to the changing international environment based on the premises of reducing unfair treatment such as U.S. intentions toward regime collapse, preemptive nuclear strike capabilities, the nuclear nonproliferation treaty (NPT), and so on.

In practice, this move toward a combined approach is realized under the framework of temporary pragmatism. Whether the statement of the Workers' Party of Korea (WPK) constitution that reads "the WPK shall accomplish the nation's independent peaceful reunification, get rid of our people's mortal enemy, the U.S. imperialists, from South Korea, reject Japanese militarism, overthrow the puppet state that is ruled by landlords, comprador capitalists and the reactionary bureaucracy, support the South Korean people's anti-U.S. and anti-puppet regime efforts, and complete the revolutionary struggle in South Korea" is still to be believed at face value is a debatable topic. The fact that North Korea took a pragmatic approach with respect to its external relations amid an unfavorable post–Cold War environment but continues to seek self-oriented socialism, independent alignment, independence under globalization, and reunification after recovering its national power or following its achievement of status as a strong and prosperous nation is suggestive in this direction.

In that sense, North Korea's efforts regarding diplomatic normalization with the United States or a peace treaty are temporary in nature, while their goals are distanced from the ultimate purpose of North Korea's advancement. The August 2001 Moscow Declaration, which addressed the removal of the U.S. military presence in South Korea as "an urgent issue that cannot be postponed in order to assure the peace and security of the Korean Peninsula and the Northeast Asian region" with an express understanding from Russia, suggests that the withdrawal of U.S. troops and the "peace" of the peninsula are still regarded as important issues for North Korea.

Likewise, North Korea's assessment of the post–Cold War era and the new international environment of the twenty-first century constitute a limited approach under partial pragmatism while largely failing to overcome the old world order. In the North Korean analysis, the twenty-first century marks a period when the willingness to engage in anti-imperialist struggle determines a people's fate as seen in the two contrasting examples of Cuba and Iraq. Under the notion that states "the nature of imperialism never changes," this dichotomous worldview points to the problems emanating from reliance on Western powers while asserting respect for independence, noninterference, and so on. While considering the United Nations an important institution, North Korea is also critical of its degeneration into a legalizing tool for a certain country to realize its strategic interests.

The current determinant for North Korea's diplomacy can be summarized as an all-out effort to establish an international environment that can sustain its military-first system. Under this framework, North Korea will strongly commit itself to a nuclear diplomacy that can guarantee its nuclear self-defense capability against the U.S. military threat, while the military-first doctrine is unlikely to be changed until the regime can sufficiently secure physical assurances of its survival from the United States and other neighboring powers. As the succession process continues and the era of a new leader under the strong and prosperous nation looms ahead, North Korea will work hard to lay down the foundations for this new age on the diplomatic front as well. In doing so, North Korea will (1) strengthen its nuclear self-defense capability in order to safeguard its state and regime integrity, and control external relations in order to secure its nuclear capability, especially vis-à-vis the United States, and win favorable grounds for its diplomatic standing as a nuclear power. Also, North Korea will (2) commence diplomatic projects that seek to stabilize the regime under the new leadership after Kim Jong-il, conduct nuclear state diplomacy to guarantee state and regime security, and carry out regional diplomacy in light of economic and national development.

The problem is that if North Korea continues to maintain its military-first diplomacy, the regime will face increased alienation and economic sanctions from its neighbors along the way. In that sense, how long can North Korea maintain its military-first rule? Especially, how durable is the economy in such a system? Will the burden of sanctions be unequally reflected in the people's economy, which could lead to political problems such as violent protests? In other words, can North Korea's military-first diplomacy continue to function in the face of long-term sanctions?

If North Korea continues to conduct its diplomacy in order to maintain and strengthen its military-first rule, its path will diverge from a twenty-first-century mode of development and it will remain a backward state. In light of a diplomatic strategy that seeks to maintain North Korea's military-first rule, the current goal sets the year 2012 as the first period when the foundation of a strong and prosperous nation will be established. In order to achieve this objective, weakening the external security threats and gaining assurances for the survival of the state, regime, and the leadership are considered to be the most important diplomatic challenges for North Korea. That said, different scenarios under a three-stage transition can be considered as in the following.

The first stage denotes the period when the nuclear program is used as an exploitable resource for North Korea to conduct military-first diplomacy. Here, North Korea will be able to consider three alternatives: obtaining recognition as a nuclear state, continuing its brinksmanship diplomacy, or completely abandoning its nuclear program. The second stage is in line with the conditions following the chosen nuclear strategy during the first stage, where regime strengthening and economic development become the primary concern. The development of the second stage will be drastically different, depending on which decisions have been made for the nuclear question. The third stage indicates a substantive materialization of the choices made during the previous stages. In this stage, the difference between the decisions of an increased backwardness that leads to greater difficulties in regime survivability and a strategic determination leading toward improved conditions will be in clear contrast.

At present, North Korea continues to use its nuclear program as an important diplomatic tool to gain security assurances from the United States and other neighboring democratic countries in the midst of a post–Cold War unipolar system led by the United States and also the unstable international environment in Northeast Asia. This situation will only allow North Korea to continue its nuclear diplomacy and its brinksmanship tactics under a limited structural environment.

For North Korea, the most favorable outcome of its nuclear diplomacy will be to achieve nuclear power status along with security guarantees and economic aid. While continuing to develop its nuclear weapons and missile capabilities, North Korea may be calculating the gains from its brinksmanship diplomacy that can bring about a de facto recognition of its nuclear power status and increase its international standing by participating in security and nuclear arms reduction talks, so that the regime can gain nuclear deterrence along with the prestige of a nuclear power state and the economic and diplomatic security benefits thereof. As such, the nuclear power status and the framework of a strong and prosperous nation will provide support for the consolidation of the post–Kim Jong-il era in North Korea. Given the fact that the new leadership will lack political and diplomatic footing for a while, inheriting North Korea's nuclear status will provide a much needed tool for the military-first rule that can sustain external diplomatic pressure. Also, the prestige of being a nuclear power state and the potential economic benefits from nuclear arms reduction talks will help to increase the effectiveness of the successor regime domestically.

If North Korea succeeds in its nuclear diplomacy and wins physical security assurances, the regime will move forward to the second stage by focusing on political and economic development of the strong and prosperous nation, and diplomatic strategies will unfold in a corresponding manner. Fast-paced reform measures and the opening of the economy will undoubtedly threaten the North Korean regime, and it will have no other choice than to maintain the *Suryong* (Dear Leader)-based controlled economy for some time. Opening up its economy and implementing internal market mechanisms in order to improve the economic efficiency of North Korea can undermine this system. Thus, North Korea would only be able to rely on selective liberalization and limited reforms for a significant amount of time. Even today, the regime is suffering from numerous problems arising from a dual economy where unofficial transactions and corrupt schemes are rampant. Even though its nuclear diplomacy succeeds, the structural limits of North Korea will lead to continued suffering from such difficulties unless there is systematic institutional reform. Moreover, the new leadership of the post–Kim Jong-il era will find itself faced by a dual challenge of strengthening internal control and increasing economic efficiency to sustain its internal legitimacy with a much weaker political standing. Therefore, the new leadership will have to oppress the unofficial sectors and increase the control of the economy while also improving its efficiency. Without reforms and liberalization, however, such goals will be extremely difficult to achieve.

During this second stage, if North Korea obtains the status of a nuclear power state, there are possibilities for the regime to extract external support by utilizing nuclear arms reduction diplomacy. However, the mix between military-first diplomacy and internal economic development will continue to pose a threat to the outside world while the amount and characteristics of the economic gains projected from potential arms reduction talks will remain limited, and the regime will find itself once more in difficulties. If North Korea abandons its nuclear programs, it will be dependent upon external humanitarian support. Nonetheless, this will ultimately diminish the regime's internal legitimacy. Therefore the new challenge for North Korea's economic development in the aftermath of a military-first, nuclear diplomacy will be contingent upon how boldly the regime can move toward reform and openness. Going through such a transition may risk the sustainability of the *Suryong*-based system for either Kim Jong-il or his successor. This will call for a change in the military-first rule to become an economy-first rule where the regime's diplomatic efforts should focus on adopting several aspects of capitalism and adhering to international norms with respect to its diplomatic strategies.

The final stage can be identified as a period when substantial changes occur in North Korea as a result of the decisions made in the previous two stages. If the regime succeeds in making a strategic decision to abandon its nuclear program in the first stage, or decides to follow the path of reform and open economy under the premise of dismantling its nuclear capabilities during the second stage, it will be able to push ahead with a diplomatic strategy tuned for advancement, albeit under different timing. This possibility will be discussed further in the next section. However, if North Korea decides to maintain its nuclear power status or continue to engage in brinksmanship diplomacy while conducting limited economic reforms, the regime will ultimately find itself unable to sustain its current structure. With its nuclear capabilities, North Korea will find more hard-line stances in diplomatic relations with its neighbors and the international community. The past sixteen years of North Korean nuclear crisis have persistently followed a pattern of extracting economic aid from the concerned parties that has contributed to regime survival while North Korea has simultaneously continued the development of its nuclear program. In between the cycle of intervention and sanctions, neighboring countries have sought both to persuade and to punish North Korea while its nuclear capabilities have continually increased. At the current stage, there has been a broad understanding among these countries that intervention toward North Korea cannot follow the form of acquiescence under the premise that the regime's nuclear development will reach an unacceptable level. Even China and Russia are approaching a fundamental judgment that intervention in North Korea cannot be maintained under continued nuclear development. Therefore, if North Korea continues to seek economic aid and internal stability from its nuclear diplomacy, there will be less opportunity for the regime to benefit from the old patterns in the distant future.

Under the status quo, little improvement in the economy will pressure Kim Jong-il or a possibly weaker successor regime of his to accept a rapidly closing window of diplomatic alternatives. Although North Korea will seek fragmented economic benefits and strengthen internal security by economically relying on China and making friendly gestures toward South Korea, the possibility of obtaining economic efficiency and garnering continued loyalty from the people, and maintaining political support from the core ruling elites including the military, will gradually decrease. At the same time, the possibility of worsening countrywide poverty, alienation from its people and subsequent defections, and the likelihood of an antiregime movement as well as disagreement and resistance among the ruling elite toward the regime's national strategy, or even signs of revolt, will increase.

As a consequence, if North Korea follows the three stages aligned with military-first rule, the opportunities for subsequent military-first diplomacy

Table 1. North Korea's Diplomatic Alternatives and Following Developments

	Stage 1	Stage 2	Stage 3
Diplomatic alternatives of North Korea	Strategic abandonment of the nuclear program	Reform and opening of the regime	Push for an economy-first, advanced diplomacy
	Continued brinkmanship diplomacy	Diplomacy toward economic aid	Reiteration of an unstable diplomacy
	Acquisition of a nuclear power status	Winning concessions through nuclear arms reduction diplomacy	Hard-line diplomacy for greater concessions
Corresponding developments	Fundamental change in military-first diplomacy	Expansion of economic aid and favorable grounds for economic opening and reform	Approach the path of North Korean advancement
	Prolonged economic sanctions	Regularized sanctions and increased economic hardship	Increased regime destabilization and a failure of diplomacy
	Impossibility of gaining de jure nuclear status	Regularized sanctions and increased economic hardship	Extreme economic hardship

will be highly limited, while its purported national objective of establishing a strong and prosperous nation will become increasingly unobtainable unless efforts are made toward adopting an economy-first, advanced diplomacy.

Gradual Decline and Search for an Exit

How would North Korea's diplomatic inclinations change as the regime decides to forgo its currently continuing strategy of military-first rule? If North Korea conducts a new type of diplomacy attuned to a normal, modern, and advanced national strategy, what would its subsequent stages and their outcomes look like? In order to achieve its goals under the norms of the international community, North Korea will have to find itself in a situation where the regime can ultimately change regardless of its temporary continuation, embrace new reforms and openings in all aspects of society, including political, economic, military, and socio-cultural sectors, and adopt an appropriate diplomatic strategy that can support other functions as well.

In a broad sense, the transition from a military-first diplomacy to an advanced, economy-first and people-first diplomacy can be understood in three phases. The first phase constitutes the period from the present to the year 2012, when North Korea fully experiences and recognizes the problems of its nuclear diplomacy and decides to abandon its nuclear program completely so that the door toward openness and reform can be unlocked. The second phase marks the gradual beginning of exchange and cooperation with the international community under limited conditions following the resolution of the nuclear issue along with normalized external relations and the procurement of necessary support toward regime transformation. At this stage, cooperation with South Korea will become more vibrant along various dimensions and issue areas. The third phase will mark the effective settlement of North Korea's advancement as the regime goes through a substantive transformation triggered by new changes in the respective sectors. In other words, this phase can be defined as a stage where North Korea seeks to become a developed country under political democratization and a properly functioning market economy and become fully incorporated into the international community. It is also the phase where the two Koreas can completely settle their hostile relations and be engaged in close exchange and cooperation toward political integration so that the prospects of a real reunification become visible. This final phase also signifies the achievement of North Korea's full diplomatic relations with the international community and its transition toward a normalized modern state in Northeast Asia.

Going back to the first phase, it is crucial for North Korea to make a strategic decision to abandon its nuclear program by gaining full assurance for the stability of the regime and the political leadership under diplomatic resolutions and having confrontational sentiments and diplomatic conflicts sufficiently resolved. This will necessitate changes in other aspects of North Korea, especially in its national strategy. The current direction toward a strong and prosperous nation under continued military-first rule considers the nuclear program an important tool to deter external security threats and gain nuclear power status at the most or at least maximize the extraction of economic aid using brinksmanship tactics in the diplomatic sector. As mentioned in the previous section, such a diplomatic alternative will essentially lead to regularized sanctions, deterioration of the economy, and weakened political legitimacy of North Korea's *Suryong* system. Thus, North Korea faces a situation where it needs to presume a realistic level of security assurances that can be addressed within the modern international arena, and maximize its functional mechanisms under the strategic decision to abandon the regime's nuclear program altogether.

As of early 2010, North Korea has been realigning its internal regime structure to establish the power succession system at an earlier pace while focusing on economic developments in order to establish a strong and prosperous nation by 2012. To gain the practical benefits necessary for the new leadership to settle in the international environment as well as in the economy, North Korea is trying to lift the imposed economic sanctions while simultaneously seeking maximum economic support by effectively utilizing the nuclear issue as a means to gain diplomatic accomplishments. In retrospect, North Korea initiated the third Kim Jong-il era through the Supreme People's

Assembly (SPA) on April 9, 2009, when it simultaneously expanded its military committee in anticipation of the succession process and increased state control. As part of the first phase, the launching of the long-range rocket and the subsequent nuclear test during this period would have consolidated North Korea's domestic political leadership. It seems that North Korea has started to seek negotiations in various ways because it reckons there has been a sufficient degree of internal realignment.

During the inter-Korean talks on April 21 and June 11, 2009, at the Gaesung Industrial Complex, North Korea seems to have shown its overall willingness to negotiate, despite its appearance of pressuring South Korea. The following massive mobilization campaigns such as the "150 day battle" and the "100 day battle" in 2009 and the November 30 currency reform suggest that the regime is pursuing its own schedule for a national strategy toward the establishment of a strong and prosperous nation. Appearances of Kim Jong-il also increased; it has been reported that there were 154 publicized events in 2009, up to December 18. This is a 77 percent increase compared with the same period in the previous year of 87 reported appearances, and also marks the highest level since the first Kim Jong-il era. On the other hand, North Korea has been attempting to change the ambiance surrounding its relationship with the United States in anticipation of the resumption of talks, while seeking economic benefits through negotiations with South Korea. It can be said that the North Korean regime will seek a consolidating diplomacy by seeking bilateral negotiations with the Obama administration as well as practical benefits in the Six-Party framework as a means to accomplish its goals to bring about the strong and prosperous nation of 2012.

Since 2010, North Korea has been exhibiting a large external strategy framework that includes both domestic policies and the nuclear program. The 2010 New Year's joint editorial pointed out that the current key issue of North Korea is to accomplish a "critical transition in people's livelihood." In order to do so, the editorial emphasizes the importance of putting a national effort into light industry and the agricultural sector. This can be understood as an effort to "increase the people's standard of living" by assigning an exceptional national priority to the development of light industry, which is reflected by its domestic circumstances in need of an increased supply of daily necessities. This is in line with the reasoning behind the effects of the currency reform, which is to boost North Korea's official economic sector by suppressing market activities where an insufficient amount of increase in the production and supply of light industry products can cause serious inflation and a revitalization of unofficial market activities (Korea Institute for National Unification 2010).

In order to solve such domestic economic hardships, nurturing a favorable international environment is critical. In the inter-Korean relations section of the 2010 New Year's joint editorial, it is stated that "opening the path to improved North-South relations is necessary. Our position to pursue better North-South relations and open up the path toward national reunification based on the historic June 15 Joint Declaration and the October 4 Declaration is firm ... [We must] seek reconciliation by putting the common interest of our people in the front seat and push for cooperative projects through exchange and contacts between different sectors of our societies. All legal and institutional hindrances need to be abolished in front of projects that serve the common public interests of the people, and free discussions and activities regarding the issue of reunification must be guaranteed at a broader level. A united people provide the critical assurance for national reunification. The Korean people in the North, the South, and overseas must unite together and form a common solidarity to advance the movement of national reunification. Facing the ten-year anniversary of the release of the June 15 Joint Declaration and the thirty-year anniversary since the proposal of forming a Confederal Republic of Koryo, the mood of reconciliation, cooperation, unity and independent reunification must be heightened at the pan-national scope."

With regard to international issues, including U.S.-DPRK relations, North Korea's approach has been mainly focused on negotiations. The 2010 joint editorial states, "[t]he fundamental problem before the peace and security of the Korean Peninsula and its surrounding regions lies within the issue of ending the hostile relationship between the DPRK and the United States. Our position remains consistent in the direction of dialogue and negotiations that can establish a firm system of peace in the Chosun peninsula and bring about denuclearization. Our party and the government shall vigorously push forward to advance good-neighborliness ripple [in a] wave of independence throughout the world, under the banner of independence, peace, and friendship."

Along with the Six-Party Talks, North Korea started to make use of a diplomatic strategy in pursuit of a peace treaty. On January 11, 2010, the North Korean Ministry of Foreign Affairs stressed the need to have a confidence-building measure between the United States and North Korea and a peace treaty in conjunction with the Six-Party Talks, following a U.S.-DPRK dialogue: "Having gone through frustration and failure, the Six-Party Talks process points out that problems cannot be solved without mutual trust among the related actors. It is our conclusion that it is necessary to pay primary attention to building confidence between the DPRK and the United States, the parties chiefly responsible for the nuclear issue, in order to bring back the process for the denuclearization of the Korean Peninsula on track. In order to build confidence between the DPRK and the United States, there needs to be a peace treaty that can end the ongoing status of war which is the original source of hostile relations. Had there been a firm peace structure in the Korean Peninsula earlier, the nuclear problem would not have emerged. Learning from the failure of the Six-Party Talks, the sequence with respect to the issue of signing the peace treaty mentioned in the September 2005 Joint Statement of Principles can be put forward as a practical demand. The DPRK Ministry of Foreign Affairs respectfully proposes holding talks among the parties directly involved in the Armistice treaty to move toward a peace treaty in this sixtieth-year anniversary of the outbreak of the Korean War. As mentioned in the Joint Statement, the meeting can be held under a separate framework or also held within the boundaries of the Six-Party Talks for the denuclearization of the Peninsula, like the current DPRK-U.S. talks that are in progress."

Put together, the strategy sought by North Korea in the current first phase can be summarized as follows. First, accelerate internal reforms to resolve the economic hardship while consolidating the succession process. Second, maximize material support from neighboring countries through reconciliatory policies. Third, improve the relationship with the United States by using the nuclear program as a bargaining chip while ameliorating sanctions and breaking away from diplomatic isolation. Fourth, nurture an environment that can provide maximum diplomatic and economic concessions from China and South Korea while participating in the Six-Party Talks.

The core issue during this phase is whether North Korea can make a strategic decision to abandon its nuclear program and seek opening and reform under the assumption that the political leadership can be maintained, or to continue seeking nuclear power status while trying to extract as much benefit as it can, given a peace treaty that can provide substantive conditions for the withdrawal of hostile U.S. policy, and, if so, under what specific conditions. Meanwhile, North Korea may also attempt to minimize the denuclearization efforts at the Six-Party Talks by linking the peace treaty issue with the nuclear negotiations and putting forward unreasonable demands in the latter so that the entire process can be stalled. In this case, the peace treaty will lose ground in terms of gaining a common understanding while the denuclearization process is left without progress, resulting in a standstill of the status quo.

So far, North Korea has been arguing for the need for a peace treaty as a precondition toward denuclearization, emphasizing its confidence-building function vis-à-vis the United States. First, the Six Parties have shared a common understanding of the need for a peace treaty. That is, they have promised that "[t]he directly related parties will negotiate a permanent peace regime on the Korean Peninsula in an appropriate separate form." Following the U.S. visit by North Korea's vice marshal Jo Myong Rok, the October 12, 2000, U.S.-DPRK Joint Communiqué suggests how North Korea had been previously materializing such an effort. The communiqué states that "the two sides agreed there are a variety of available means, including Four Party talks, to reduce tension on the Korean Peninsula and formally end the Korean War by replacing the 1953 Armistice Agreement with permanent peace arrangements," and "[a]s a crucial first step, the two sides stated that neither government would have hostile intent toward the other and confirmed the commitment of both governments to make every effort in the future to build a new relationship free from past enmity ··· the two sides reaffirmed that their relations should be based on the principles of respect for each other's sovereignty and noninterference in each other's internal affairs, and noted the value of regular diplomatic contacts, bilaterally and in broader fora ··· agreed that resolution of the missile issue would make an essential contribution to a fundamentally improved relationship between them and to peace and security in the Asia-Pacific region."

On July 22, 2005, the North Korean Ministry of Foreign Affairs provided a similar comment by the spokesperson under the title, "DPRK Ministry of Foreign Affairs Spokesperson's Remark on the Argument for the United States to Change Its Policy to Maintain the Armistice Regime into a Policy toward the Establishment of a Peace Regime." It states, "The transition from a cease-fire situation into a peace regime will eliminate the U.S. hostile policy toward the DPRK and its nuclear threat and lead naturally toward the realization of denuclearization. As a result, the establishment of a peace regime constitutes an absolutely necessary step to achieve the goal of denuclearization in the Peninsula. The transition from an unstable armistice regime into a firm peace regime will turn out to be a peaceful process to resolve the problems of the Peninsula. Therefore the process regarding the establishment of a peace regime must be carried out in a way that contributes to the nurturing of an environment that contributes to the realization of the peaceful coexistence between the DPRK and the United States, and the peaceful reunification between the North and the South. The successful initiation of the peace process in the Peninsula will not only contribute to the peace and security of the Korean people, the Northeast Asian region and the world, but also become a critical impetus to resolve the nuclear problem in the forthcoming Six-Party Talks."

Such remarks indicate that North Korea's concern over U.S. hostilities is consistent and that it considers the peace treaty to be a prerequisite for denuclearization. If so, in order to induce North Korea's decision to abandon its nuclear program in the first phase, questions such as how to solve the problems regarding the peace treaty, the issue of North Korean stability, and denuclearization need to be addressed.

The foremost goal for North Korea is to lead the Six-Party Talks in a way that is closest to the objectives of its national strategy, and to secure as many preconditions as possible. Although there are observations that North Korea has already made its strategic decision to become a nuclear-power state, there is no certain indicator to accurately assess the change in its strategy regarding dismantlement. For now, it would be more precise to say that whether Kim Jong-il is only seeking practical benefits such as economic support, break-away from diplomatic isolation, and lifting sanctions while leaving behind the nuclear issue to his successor is largely unknown. The transference of the nuclear issue under an uncertain environment where his successor's domestic-political and external-strategic capabilities are untested and also without knowing whether the military and government elites will support the succession process despite sanctions is a highly sensitive issue. Therefore, from Kim Jong-il's perspective, it can be hypothesized that nuclear dismantlement is a considerable possibility if the U.S. hostilities are eliminated entirely, along with guarantees for the succession regime to enjoy diplomatic and military assurances against security threats.

Regardless of the strategic decision to become a nuclear power or to consider dismantlement depending on the negotiation results with the United States, North Korea will try to gain as many diplomatic and economic benefits as possible through its relations with China and South Korea. Toward China, Pyongyang will seek to provide a detailed explanation of the regime's current position and gain understanding for its recalcitrance over the nuclear issue as a result of hostile U.S. policies or any passive approach thereof. Accordingly, the regime will attempt to secure its position to be able to receive economic aid from China even though the process of dismantlement is delayed. Even when the Six-Party Talks resume, North Korea will consider the possibility where China would stress its function as a diplomatic intermediary as an impartial party while keeping the regime under its sphere of influence as problems arise within U.S. policy toward the North.

As for South Korea, it is likely that North Korea would take a reconciliatory approach by emphasizing inter-Korean economic cooperation while restraining its diplomatic and military voices in either the Six-Party Talks or the peace treaty dialogue. If the regime focuses on maximizing practical benefits while providing little recognition of South Korea's position, there will be little to lose for North Korea even if the Six-Party Talks fall apart once more.

Such reasoning should be seen as North Korea's intention to acquire favorable grounds at the peace treaty negotiations by increasing prerequisites and raising the entry barrier for South Korea.

It is useful to consider the implications of North Korea's intentions with respect to its nuclear program as being both aggressive, such as building the strong and prosperous nation, and defensive, such as strengthening its deterrence and creating an environment that can bring about economic development. In the latter case, North Korea will ultimately give up its nuclear program as it fails to gain recognition as a nuclear-power state by winning satisfactory concessions for its regime security and simultaneously gaining large-scale support for economic development.

In doing so, it is important for North Korea to gain an acceptable amount of material security guarantees. This is in line with North Korea's desire to have specific physical and institutional guarantees that can back up promises that the United States and its allies, South Korea and Japan, will not engage in any kind of threatening actions. In this respect, the United States' deliberation on North Korea's demands and assurances—including the signing of the peace treaty and withdrawal of U.S. troops in South Korea, the complete denuclearization of the Korean Peninsula along with U.S.-DPRK mutual inspections, and even the withdrawal of the U.S. nuclear bases from the entire Northeast Asian region—will be an important factor. However, such forms of security assurance will be nearly impossible to provide, given U.S. global strategy and interests that relate to the Korean Peninsula as well as to Northeast Asia. Therefore, only by accommodating the United States with respect to security assurances or by increasing the level of confidence between the United States and North Korea through close bilateral talks will it be possible to reach an agreement that is mutually satisfactory.

Under a modern international political system that is organized under the basis of anarchy, and while security assurance mechanisms consist of physical and institutional aspects, the United States will be able to provide North Korea with common interests, a rightful position in Northeast Asia, and confidence building. Being entangled with fragmented security assurance measures can be thought of as providing a simplistic response to the current complex international political arena. Instead, North Korea should assess how much of its interests are aligned with its neighbors, build more confident relations by winning over cooperative groups, and improve its diplomatic strategy so that neighboring countries realize North Korea's existence can bring benefits to Northeast Asia.

Domestically, there should be coexisting or conflicting voices between aggressive and defensive nuclear strategy, or between camps that strategize a military-first, strong and prosperous nation with nuclear capabilities and that consider a strategic decision to abandon the nation's nuclear programs and transition into an economy-first era. In order to induce North Korea to make the strategic decision to relinquish its nuclear program and adopt a different diplomatic alignment, it is necessary to send various delicate signals to those who are in support of the idea in North Korea that the political leadership and the regime will be preserved in doing so, while this transformation will also be helpful for the development of North Korea. While being too up-front with a promise of regime security might easily gain criticism from hard-liners in neighboring countries, being anxious about North Korea's aggressive responses and devising an overly conservative approach in North Korean policy is also problematic. Instead, it is necessary to continue to present the direction of North Korea's advanced diplomacy while providing assurances for the willingness to carry out relevant actions through multiple channels.

Ultimately, the objective that must be accomplished along with North Korea's national and diplomatic strategies is the coevolution of neighboring countries' strategy toward North Korea. That is, both internal and external diplomatic coevolution is needed. In doing so, it will be important to get over the existing perspective that considers North Korea's nuclear program as an issue of proliferation that can essentially be resolved through a regime change and recognize that it is directly connected to the issue of the regime's political survival. Solving the nuclear program necessitates a comprehensive approach to the issues revolving around North Korea, which requires an understanding of not just one's own North Korean policy but also the North's broader Northeast Asian policy in tandem, expanding to issues such as the North Korean anxiety over regime survival and the future positioning of North Korea in the regional setting to come.

Coevolution Strategy for an Advanced North Korean Diplomacy

Implementation and Reform

North Korea

Once the security problems are resolved, North Korea's progress to the second stage marks a period of transition for the regime. While receiving external support, North Korea's military-first rule will gradually transform into an economy-first system, and the regime will seek coexistence with the international community. During this process, North Korea will accomplish an overall change in its domestic settings, including political, economic, and social conditions thereof.

An important operational objective would be to support this process diplomatically and create a virtuous circle that draws North Korea's national strategy toward an economy-first and people-first agenda as external relations improve. During the transition process, North Korea will need external economic support in order to achieve an economy-first system, while the issue of securing assurances for regime stability during the transition will also come to the fore.

The imminent task for North Korea would be to establish normal diplomatic relations with its neighboring countries and sign security and peace assurance treaties to bring about a system of peace, based on nuclear dismantlement and improved external relations. Based on the subsequent accomplishments of dismantlement, it would be desirable for North Korea to rejoin WMD (weapons of mass destruction)-related international regimes such as the NPT (Nuclear Non-Proliferation Treaty) and the IAEA (International Atomic Energy Agency). Once it is free from its neighbors, and the international community's suspicions regarding illicit production and export of WMD, North Korea will be able to substantially alleviate concerns regarding potential redirecting of its acquired large-scale economic support into illicit external activities. Afterward, it would be favorable for North Korea to accept strict inspections and monitoring from the international community and to adhere to international norms faithfully.

At the same time, neighboring countries also need to work together in order to relieve North Korea's diplomatic and political anxieties following denuclearization. In due course, there will be a sufficient amount of compensation through the energy supply as well as a sequenced lifting of sanctions related to WMD, in response to nuclear abandonment. Acquiring a diplomatic status that allows full-scale economic support would follow accordingly.

Once dismantlement is complete and the peace process in Northeast Asia takes off, North Korea will find the opportunity to guarantee its regime survival and development through diplomatic relations. Instead of unlikely requests that North Korea would make in order to establish a peace system, including but not limited to the withdrawal of the U.S. troops in South Korea and a mutual inter-Korean inspections regime, all the relevant parties would come together and build an international control regime that can prevent any violent confrontational behavior on the Korean Peninsula, based on political and military confidence building. To the extent that the diplomatic latitude of the Korean Peninsula is unharmed, this is linked to a Northeast Asian multilateral security cooperation mechanism that will ultimately guarantee North Korea's regime security.

During this process, North Korea will be able to further its relations with traditional allies and friendly countries such as China and Russia, while simultaneously normalizing diplomatic relations with the United States, Japan, and other countries. Although the continuance of North Korea's alliance with China is a matter to be determined through bilateral discussions, the fact is that the probability of a continued cooperative relationship between the United States and China exists, and once the risk of having another war on the Korean Peninsula is eliminated, the DPRK-China alliance will be reconfigured into a strategic partnership that can guarantee regional security around the Korean Peninsula. This is on the same track of expanding the function of the ROK-U.S. and U.S.-Japan alliances to a regional role. The fact that twenty-first-century alliances are no longer adhering to a traditional framework that defined a certain enemy or a security threat, and instead are transforming into a cooperative system that expands to mutual joint responses against transnational and human security issues, the bilateral relationship between North Korea and China is likely to transform into a new functional role. Russia is also considering opportunities in economic and energy cooperation with North Korea, such as oil, gas, and railways, which can serve its own as well as the North's and the entire peninsular interests once the nuclear issue is resolved, suggesting that there will be an accelerated economic cooperation between the two.

The most significant issue during this stage is the normalization of diplomatic relations between North Korea and the United States. Once hostile relations cease to exist under the peace process following North Korea's denuclearization during the early phase of the second stage, normal diplomatic relations can come about in a shorter period of time. This critical improvement would allow U.S. investments to flow into North Korea by various social entrepreneurs and companies, and their social and cultural activities will rapidly reduce the sense of the U.S. threat while also contributing to the furthering of their bilateral relationship. North Korea will also be able to actively participate in the creation of a regional peace regime including Northeast Asia, while DPRK-Japan diplomatic relations can be brought about as well.

Following diplomatic normalizations, it is possible to expect that a gradual intensification of economic interdependence would follow. Through consultations, issues such as the Japanese reparations payments and the abduction cases can be resolved and the DPRK-Japan relationship can develop in a new dimension as Japan actively invests in North Korea. For Japan, strengthening the relationship with North Korea will serve its strategic interest in reaching out into the Asian continent under a growing recognition of the need to develop a more active Northeast Asian policy, added to the cooperative relationship with the United States as its diplomatic strategy.

As North Korea's external relations normalize, and pressures such as the economic sanctions are alleviated, it will be desirable for the regime to fully enter the path of opening and reform. However, the North Korean political leadership may try to control the content and the pace of these developments as they fear they may become a threat. Concurrently, North Korea will also feel the need to increase the regime's economic efficiency as the process continues. While helping North Korea to effectively combine these two seemingly confrontational tasks by strengthening economic aid and support, neighboring countries need to be cautious about criticism and intervention in North Korea's political system so that it can maintain its integrity and continuity.

Not only would North Korea receive bilateral support from its neighboring countries, but it will also be able to enjoy greater multilateral support from international organizations or consortia in its support. To open up and reform the system, North Korea should embrace capital investments, join international organizations, and bring about a fundamental change in the financial, infrastructure, and education sectors that are critical for implementing a market economy. At the same time, North Korea's diplomatic efforts in international fora need to be more active as well.

Likewise, North Korea will be integrated into the Northeast Asian system as a normal state in the second stage, with a steadily growing economy that is not threatening to its neighbors. Such a change will incur significant political and economic changes, which creates a risk that the existing order under the *Suryong* system will collapse. Therefore, North Korea should try to maximize international support for a reformist *Suryong* system through diplomatic endeavors and work toward gaining diplomatic and economic support to strengthen its regime. A broad recognition that it will be better for both North Korea and its neighbors for the regime to go through a gradual transition from the current *Suryong* system to a reformed, economy-first system than going through sudden changes owing to rapid economic restructuring as well as political confusion is important. As market factors develop and social and cultural exchanges with neighboring countries expand, North Korea will need to retain as much of its personnel, economic, and cultural resources as possible so that it can mature into an advanced country in Northeast Asia.

The role of nonstate actors will rapidly become more important during the economic reforms in North Korea. As seen in socialist transition countries such as China and Vietnam, the role of companies, individuals, and other economic interest groups will become important while market factors gradually expand. Such a change will also bring much change to the diplomatic landscape, which will trigger the transition into network-based diplomatic relations that can connect many actors in between state-to-state or government-to-government relationships. Therefore, even though the government controls the overall diplomatic process, it will be better to adopt a diplomatic strategy that fosters the connectivity of various actors throughout the network so that it can contribute to the long-term development of North Korea.

Inter-Korean relations will also become a key factor in North Korea's economic development and political stability. South Korea obviously has the largest interest and stake in North Korea's economic development, with a strong willingness to invest in the country. As North Korea, being confident about its political system, furthers inter-Korean exchange, South Korea should also get rid of aggressive North Korean strategies, such as absorption-based reunification, and adopt a strategic mindset to promote long-term peaceful coexistence and economic and socio-cultural cooperation with the North. As North Korea gradually changes by opening up and taking reform measures, investment, trade, and personnel exchanges between the two Koreas will expand further while the commonality between the two peoples will naturally increase. It would be best for North Korea to have its transitional system naturally coexist with South Korea's market economy by controlling its speed and promoting contact and exchange between the two societies.

South Korea and the International Community

In order for North Korea to become an advanced modernized state through opening up and gradual reform that is in line with the current direction of the international community, a parallel effort must be made by its surrounding neighbors under the notion of coevolution. In doing so, North Korea needs to make a strategic decision to abandon its nuclear program and implement policies that are necessary for its execution, while other countries work together to bring an end to the North Korean nuclear crisis so that external economic support can be expanded along with the signing of the peace treaty for the Korean Peninsula that can improve the regime's relations vis-à-vis its neighbors. It is true that the current Six-Party framework has a functional limit in its institutional structure, which seeks the resolution of the North Korean nuclear crisis.

Therefore it is necessary for North Korea to actively participate in building a network of dialogues to settle the crisis and ultimately overcome the nuclear issue through various multilateral talks along with bilateral talks with the United States as well as with South Korea.

In an effort to accelerate North Korea's economic cooperation backed by diplomatic relations with influential countries in the early part of the twenty-first century in Northeast Asia, the United States and Japan both need to work toward normalization of relations, invigoration of trade and investment, and promotion of personnel as well as material exchanges vis-à-vis North Korea. The leading members of the regime and its people need to establish political and economic institutions as well as an ideological framework that will be in keeping with global standards by actively visiting neighboring countries, which can be facilitated through various exchange programs, track-two strategic dialogues, and educational programs for youth.

During the process, it will be important to carefully avoid jeopardizing the North Korean regime and not to incite factions that can bring about a contingency under hasty reforms and opening up. It would not be beneficial for North Korea nor for its neighbors to face a situation where the process toward change leads to political instability and greater security risks. There can be many trials and errors along North Korea's transition to a market economy. Thus, it will be necessary to support North Korea's switch to an economy-first system under a reformist *Suryong* structure that is conducive to stable economic development. That is why neighboring countries adopting market economy and free democracy need to provide various educational programs that can help North Korea's economic actors to quickly learn about their operational mechanisms and promote them through policy implementation.

International support would also be an important factor to induce reform and opening of North Korea. Given the fact that the regime lacks basic infrastructure in areas such as energy, finance, education, and social welfare, providing a suitable economic platform will not be an easy task. Thus, it is desirable to develop a program such as the forming of a transition support team that will center international organizations to facilitate North Korea's reform and opening up while stabilizing its economy in the short run.

The strategic dialogue on dismantlement and the resolution of North Korea's nuclear issue is bound to converge in a Northeast Asian regional strategy that sits at the forefront of North Korea strategies of the neighboring countries as well as North Korea's position in the region. Having said that, North Korea will require new diplomatic strategy and philosophy conducive to agenda development as well as international leadership within the Northeast Asian framework that can be channeled through the dialogues on the nuclear issue. In doing so, North Korea can look into its record of diplomacy and actively develop values that can be accepted by the current and future international regimes and form an aggressive strategy that can contribute to peace on the Korean Peninsula and in Northeast Asia. During the process, various forms of peace regimes can be brought about and relations with neighboring countries can improve so that a momentum for economic development can also be spurred through opportunities such as the Japanese reparations fund. International financial support, personnel training and education necessary for economic development, and other forms of assistance can be provided as well.

South Korea should also avoid a strategy that seeks reunification by absorption that could occur if the government attempts to excessively intervene in the reform and opening up process in North Korea. That fact that the negative consequences of a rapid reunification process can be prevented should not be overlooked, whereby improving inter-Korean relations and establishing a network connection throughout the Korean Peninsula can be realized by North Korea's self-controlled transition to an economy-first system. Hence, South Korea should establish a clear vision for the future governance system of the Korean Peninsula prior to the transition into the second stage, and succeed in its coevolution strategy that is compatible with North Korea's evolution.

Transformation

North Korea

In the third stage, North Korean diplomacy should complete the virtuous circle that is beneficial for its neighbors as well as for the region and the entire world in conjunction with its effort to maximize national capacity. A successful switch from military-first to economy-first rule will bring about qualitative differences in all aspects of North Korean society, including politics, the economy, socio-cultural aspects, human rights, and so on. When the process of democratization and the transition toward a market economy inclusive of the global system occur and the socio-cultural as well as the human rights situation advance gradually, North Korean diplomacy will function as a vehicle for such changes while also acting as a gravitational force that leads the transition.

As a result, North Korea will be reborn as a middle-sized Northeast Asian state with a population of about 30 million. Despite its past record of bearing negative labels such as an evasive state or an "axis of evil," and suffering from the hardships stemming from its military-first politics, the new North Korean state will have the reputation of a case where the regime took an independent track toward democracy and economic development through a successful transformation of its institutions. During this process North Korea would not only be able to achieve an effective democracy and market economy but also be able to conduct its diplomacy in a way that can contribute to a peaceful transition against the uncertainties revolving around the region by strengthening strategic bilateral developments with countries such as the United States and China within the Northeast Asian international political environment. It can also be expected that North Korea can become an exemplary state for third world and even the impoverished fourth world countries with respect to its development model and expand its sphere of influence by establishing ties with them.

North Korea will be able to break away from its defensive posture in this stage and conduct a more aggressive and active diplomacy that is beneficial for itself as well as for its neighboring countries based on its newly acquired capacities. While leading its peace regime vis-à-vis other countries in the military security area, North Korea will be able to project its own diplomatic strategy and also enjoy a greater opportunity to participate in various subject matters such as human security or a comprehensive security regime. In cooperative multilateral regimes in Northeast Asia as well as in East Asia, North

Korea will also be able to function as a middleman between China, Russia, South Korea, and Japan, while consolidating its position through strategic relations with nearby powers.

Through close cooperation, North Korea will be able to play a more active role in promoting gradual integration by overcoming the divided status of the Korea Peninsula. In the unpredictable regional environment of Northeast Asia that can be addressed by either a balance or a conflict of power, North Korea will also find possibilities to contribute to the settlement of peace in the region. At this stage, North Korean civil society will also enjoy much development so that inter-Korean relations can deepen by personnel and cultural exchange through a civil society network between the two Koreas. If the growth of civil society in North Korea can converge to global standards, the regime's model of economic development and political democratization can be exported globally to other regions beyond Northeast Asia.

In the economic sector, North Korea's diplomacy can deepen mutual interdependence with its neighboring countries and form a complex free trade regime, moving forward to an integrated Northeast Asian economy that connects each country including the United States, South Korea, China, Japan, and Taiwan. Such an economic integration will provide an important foundation for the regional integration of Northeast Asia, whereby its success can induce an inter-Korean integration process that is not threatening to North Korea. The regime will be able to rid itself of its former position as a weak economic state and devise its own development model to conduct active exchanges with the third world based on postcommunist transition and the transformation from an evasive state to a normal state, and conduct a contributing diplomacy to its partners in the process.

North Korea's change in the third stage can be helpful to bring about liberal transformations in Northeast Asia, such as democratic peace, peace through markets, institutional peace, and so on. Although such liberal aspects do not automatically guarantee peace, the likelihood of having a transnational civil society through democratization, expanded regional public sectors, and an information-knowledge-based community will increase. Mutual close economic integration will likely occur through market mechanisms such as the deepening of free trade regimes, financial cooperation, and mutual investment relationships. In that sense, North Korea's effort toward advancement will contribute to the creation of a multilateral regional cooperative structure that will manage a peaceful transition of power in Northeast Asia.

Such directions for North Korea's diplomacy will assist the changing power composition in Northeast Asia of the early twenty-first century in an orderly, institutionalized manner. Also, North Korea will be able to conduct a mediating role that will help prevent excessive competition between regional powers and promote multilateral cooperation as a medium power in Northeast Asia.

South Korea and the International Community

Along with North Korea's transformation, its neighbors will also be able to work toward coevolution in ways that can provide support for the regime's integration in the region as well as in the world. By establishing its legitimate status in Northeast Asia, North Korea's economic and socio-cultural influence will expand beyond the region. It will be able to strengthen its global influence by proposing a North Korean success model marked by a transformation from a failed state to a successful state and from a modern communist totalitarian state to a twenty-first-century advanced state.

As the process of transformation unfolds, the United States will be able to establish a strategic cooperative relationship with North Korea and discuss future stability in the military and security aspects of Northeast Asia together, while simultaneously forming closer mutual interdependence between the two economies. By actively participating in the procedures within international organizations and the U.S. effort to resolve global issues such as terrorism, the environment, and the illegal drug trade, North Korea will be able to overcome its past evasive conduct and provide its services to global human security issues as well. Regional countries and international organizations should support such efforts by North Korea to reconfigure its role in the international community. Relations with China can also advance into a twenty-first-century alliance that can provide practical benefits to North Korea under dual strategic relations vis-à-vis the United States and China. Of course, the changing U.S.-

China relations toward the mid-twenty-first century will be an important subject for not only North Korea but for other countries as well.

During this process, North Korea will be able to establish its relations with regional neighbors that are conducive to multiple strategic cooperation structures, while neighboring countries will also need to recognize North Korea's active role in the region and work toward establishing relations vis-à-vis North Korea.

Inter-Korean relations will also be an important variable. At this point, there will be much less need for South Korea to assert the goal of formal reunification with North Korea, because the two institutions will already be extensively harmonized. If the two Koreas can form a de facto economic and socio-cultural community, political integration can be initiated from a long-term perspective and recognition of that as a natural process. At the same time, if the two Koreas can participate together in the regional integration process as the movement emerges, both will be able to form a multidimensional, multi-issued, network-based cooperation that can lead to a new future cooperative system of governance on the Korean Peninsula. In this respect, the two Koreas need to work together to develop and promote such a cooperative relationship with each other. ■

Table 2. North Korea's Diplomatic Alternatives and Following Developments

	North Korea's Diplomatic Strategy	Coevolution Strategy of Neighboring Countries
1st Phase of Advancement	Strategic decision on nuclear disarmament Diplomacy for the regime security guarantee Increase in economic aid and diplomatic normalization	Diplomacy to resolve the North Korean nuclear issue by urging the abandonment of the nuclear program Diplomacy to guarantee the stability of the North Korean system and regime
2nd Phase of Advancement	Transformational diplomacy to economy-first system, based on security guarantee Deepening of economic interdependence with states in Northeast Asia and diplomacy for sustainable reformist *Suryong* regime Diplomacy to expand a peace regime on the Korean Peninsula and in Northeast Asia	Diplomacy to support North Korea's transformation to an economy-first system and reformative *Suryong* system Diplomacy connecting peace on the Korean Peninsula, stability in the region, and the transformational diplomacy of North Korea
3rd Phase of Advancement	Securing North Korea's status in Northeast Asia and in the international community Diplomacy to increase Pyongyang's influence by actively participating in democracy, market economy, and institutional peace in Northeast Asia	Diplomacy to pursue active relations with North Korea so that Pyongyang could transform from a failed state into an advanced state in the twenty-first century

■ References

Chun, Duk-Sung. 2004. *Seon-gun-jeong-chi-e daehan i-hae* [Understanding of Military-first Politics]. Pyongyang: Pyongyang Press.

Eom, Gukhyeon, and Geumcheol Yun. 2006. *Jo-seon-ban-do pyeonghwa-bo-jangmun-je* [Issue of Peace Keeping on the Korean Peninsula]. Pyongyang: Pyongyang Publication.

Government Statement of Democratic People's Republic of Korea. 2003. "Haeg-mu-gi-jeon-pa-bangji-jo-yag-eu-lo-bu-teo tal-toe" [Withdrawal from the Non-Proliferation Treaty]. Korean Central News Agency. January 10.

Kang, Sung-Gil. 2002. *Seon-gun-si-dae-ui jo-gug-eul ga-da* [Going to the Homeland of the Military-first Era]. Pyongyang: Pyongyang Publication.

Kim, Gye-Dong. 2002. *Bug-han-ui oe-gyo-jeongchaeg* [North Korea's Foreign Policy]. Seoul: Baeksanseodang.

Korea Institute for National Unification. 2010. "2010 nyeon sin-nyeon gongdongsa-seol bun-seog" [Interpreting North Korea's Joint New Year's Editorial]. On-Line Series, 10-01.

Korean Association of North Korean Studies. 2006. *Bug-han-ui tong-il-oe-gyo* [North Korea's Unification Diplomacy]. Seoul: Gyeonginmunhwasa.

Korean Central News Agency. 2000. "Jo-seon-min-ju-ju-ui-in-min-gonghwa-gug-gwa mi-hab-junggug sa-i-ui gongdongkommyu-ni-ke" [DPRK-the U.S. Joint Communiqué]. October 12.

Paik, Hak Soon. 2003. "Dae-namjeon-lyag" [South Korea Policy]. *Bug-han-ui gug-ga-jeon-lyag* [North Korea's National Strategy]. Center for North Korean Studies at the Sejong Institute. ed. Seoul: Hanul Academy.

Rodong Sinmun, 2002. "New Year's Joint Editorial." January 1.

Statement by the Spokesperson of North Korea's Ministry of Foreign Affairs. 2002. "Jo-mi sa-i-ui bul-ga-chimjo-yag-che-gyeol-i haeg-mun-je-hae-gyeol-ui hab-li-jeog-i-go hyeon-sil-jeog-in bangdo" [Concluding non-aggression treaty between DPRK and the U.S. is a reasonable and realistic measure to resolve the nuclear problem]. Korean Central News Agency. October 25.

CHAPTER 4

Getting Out of the Military-First Dilemmas

In Search of North Korea's Coevolution Military Strategy

Jihwan Hwang

Introduction

EVER SINCE KIM IL-SUNG'S DEATH IN 1994, THE MILITARY-FIRST POLICY has been maintained as Kim Jong-il's principal strategy for governance and survival in North Korea. The policy's effectiveness as a force in domestic politics and political propaganda is debatable, but its critical flaw is that it creates doubts as to whether it will be effective enough to ensure North Korea's security and powerful state building in a rapidly changing world order.

The military-first policy is designed to enable North Korea to become a strong and prosperous state by 2012, but in fact it is restricting the development of nonmilitary sectors. Running the state through the army and focusing almost all of state capacity on the military disrupts the effective distribution of resources and aggravates the problem of an overweening military and an excessive emphasis on security. Thus, despite its designers' intentions, the military-first policy will inevitably fail as a national strategy.

Kim Jong-il initially adopted this strategy in order to facilitate the maintenance of his regime. There is a possibility that despite Kim Jong-il's being aware of the policy's imminent failure, a lack of alternatives is forcing him to hold on to it. Insistence on this strategy will foster not the construction of a strong and prosperous state but both external and internal ineffectiveness, weakening both the regime and the state, and eventually bring them down. On the other hand, if the current strategy is abandoned and North Korea makes new strategic decisions, its chances of survival will be greater and it will certainly be able to become a normal state in the international community.

This chapter points out the structural problems that North Korea's military-first policy entails and proposes a new direction for development in a so-called coevolution strategy. Pyongyang could adopt it by modifying the military sector, aiming to transform North Korea from both the inside and the outside. The North Korean issue is not one that can be solved simply by changing the international community or the South Korean government's North Korea policy. The problem with President Lee Myung-bak's Grand Bargain or the Obama administration's "Comprehensive Package" is that they lack programs that would transform the North Korean state itself. Meanwhile, due to the current domestic situation and the political environment on the Korean Peninsula, the possibility of the problem's being successfully resolved by North Korea's leadership by promoting a different autonomous reform program in the Chinese or Vietnamese style is rather low. In the beginning stages of reform there would be heightened socioeconomic instability due to the loosened grip on internal politics, and the leaders would feel increasingly threatened by their relative weakness compared to their strong South Korean neighbor. Therefore, for the successful survival of North Korea, there must be a coevolution strategy consisting of giving up its nuclear weapons and promoting an autonomous reform program, while its neighboring states simultaneously ensure and support its safety.

The strategy proposed in this paper is not a completely new concept. Despite certain defects, former president Kim Dae-jung's "Sunshine Policy" and President Lee Myung-bak's "Denuclearization-Openness-3,000 Initiative" both aim for internal change in North Korea and simultaneous support from the international community. The discussions on the Korean Peninsula's peace system during the Six-Party Talks follow the same context, aiming to shape North Korea's strategy by changing the security environment of the peninsula. However, with the absence of mutual trust, such an approach is bound to take on a limited form, and a more fundamental strategy that satisfies both sides is required.

Advocating both guaranteeing the system and giving up nuclear weapons may seem reckless and unrealistic, but the current environment surrounding the Korean Peninsula is actually quite favorable for resolving the North Korean issue. This situation becomes clearer if we understand that North Korea is facing a threefold problem of nuclear weapons, economic crisis, and the succession of its leader. It also requires a fundamental reform in all other sectors, including politics, international relations, the economy, and the socio-cultural realm (Chun 2009). In order to achieve fundamental reform, the military-first policy must be abandoned and a new strategy needs to be devised. But given North Korea's political situation, unless change in the political structure comes first, there is a very low chance that nuclear weapons will be given up and economic reforms will be carried out. In this context, a coevolution strategy aims to make North Korea decide its strategy regarding nuclear weapons and economic reforms during its process of achieving a stable succession. Such a strategy aims for an eventual shift from a military-first policy to an economy-first policy.

Coevolution strategy in the military sector aims to resolve the problem of "excess security" and excess caused by the military-first policy, the essence of which is giving up nuclear weapons. Just as in other sectors, it is comprised of three stages. The first stage consists of North Korea declaring the shutdown of its nuclear weapons program in the process of the succession while the international community ensures external and internal security. The second stage is North Korea actually carrying out what it has promised and shaking off the security dilemma by moving from a military-first policy to an economy-first policy. The final stage results in the reduction of armaments of both South and North Korea, the latter participating in East Asian multilateral security cooperation after having adopted an entirely new approach to becoming a strong and prosperous state.

Understanding and Evaluating Military Strategy in the Military-First Era

Military Strategy in the Military-First Era

Kim Jong-il's military-first policy can be interpreted as having succeeded and developed Kim Il-sung's idea of prioritizing the military and applying this to national strategy during the era of armed struggle (Kang 2002, 17). Kim Jong-il said that "safeguarding and persisting in our socialism and completing the great achievement of revolution by using force is our Party's unwavering will and conviction." He added that he was not willing to pursue any changes, emphasizing that he had "reached a firm decision to overcome any hardship by relying on the strength of the gun" (Jeon 2004, 15−16). However, it is said that Kim Jong-il's policy is different from Kim Il-sung's strategy to a certain extent, because there was formerly no party, state, or formal military during the process of nation-building and Kim Il-sung was mostly dependent on the proletariat class when conducting warfare against the Japanese. On the other hand, in a situation in which both a party and a state exist, Kim Jong-il's military-first strategy insists on relying on strong military power represented by the North Korean People's Army to protect socialism and face the international community's hostile policies (Jeon 2004, 16−17).

This aspect is made evident in the "military before labor (先軍後勞)" terminology emphasized by the military-first policy. It is based on the notion that in the fulfillment of socialism the proletariat force can carry out its role only with the support of strong military power because even the proletariat force, which is the spearhead of the struggle for socialism, cannot escape the fate of getting enslaved if there is no strong military (Kang 2002, 22–30). Thus this perspective places more importance on the revolutionary spirit of the military rather than on that of the proletariat class and regards the military not as a means but as the leading force of politics. This reflects how, with the end of the Cold War and the collapse of socialism, North Korea's internal and external security dilemma has intensified and the proportion that the military sector takes in national strategy has increased accordingly.

In the end, there are two types of military-first policies: the "political strategy placing the military as the top priority of national affairs" and the "political strategy having the military as the core and main force" (Jeon 2004, 19; Kang 2002, 19–22; Eom and Yun 2006, 150–151). In the perspective of military strategy, the former is for foreign security and the latter for internal stability of the regime.

Military Strategy of the Military-First Era Aiming for External Security

North Korea claims that military-first policy is the "political strategy placing the military as the top priority in national affairs" (Jeon 2004, 20–26). Because, it says, the international community, including the United, States, is menacing its right to exist, it is inevitable that the North Korean military sector be treated as of the utmost importance in order to be able to defend the state's other duties in areas such as politics, economics, and culture. The existing traditional socialist theory was perceived to be based on a materialist view of history aiming to construct socialism with improvements in the economic sector, such as in production, followed by the strengthening of the military sector. In other words, the military was thought to be reliant on the economic sector. North Korea asserts that a military-first policy works based on creative principles that differ from traditional socialist theory because it proposes first and foremost the strengthening of the military based on *Juche* ideology (Jeon 2004, 21–23).

Because the military is perceived as the most important factor, the military-first policy devotes itself primarily to strengthening national defense. For this purpose, it is especially important to focus on developing the industrial sector of national defense because it is essential for establishing a defense system entirely for the people and the state, which would thus increase national defense capabilities. In addition North Korea emphasizes that unlike imperialism and an arms race, this type of politics aims to contain war and attain peace and not to threaten the other states. Its goal is to become an invincible military power that is able to secure the state's and nation's rights to life and sovereignty and guarantee the construction of a strong and prosperous state (Eom and Yun 2006, 156–167).

North Korea's external security and military strategy can be interpreted as so-called military-first pacifism (Hwang 2009, 119–120). It is the claim that North Korea, by becoming a strong and prosperous state through its military-first policy, gains deterrent power against the United States and consequently maintains peace in regions surrounding the Korean Peninsula. This holds the basic underlying assumption that the main source of threat in the peninsula is not itself but the United States. Especially on the issue of nuclear weapons, North Korea maintains that the problem originated not from North Korea's nuclear weapons program but from the United States' nuclear policy, which brought the weapon to the peninsula in the first place, and the hostile attitude toward the North during the Cold War.[1] Thus it is most important for North Korea to acquire a balance of power against the American threat and end antagonistic policies to ensure its safety. The North Korean belief that military power is the only possible way to attain external security and prevent war is in line with the realism in international relations theory, which states that a balance of power may contain war and make peace possible (Woo 2007).

Thus military-first policy requires the production and possession of missiles and nuclear weapons, because North Korea sees these as the most effective way of national self-defense against the American threat. It accepts the so-called balance of terror, that there has been no war between nuclear powers (Eom and Yun 2006, 160).

Strategy of the Military-First Era Aiming for Internal Stability of the Regime

The reason why North Korea defines military-first policy as "a way of placing the military as the core and driving force of politics" is to pursue the stability of the system by emphasizing the military sector, which is most essential for the maintenance of the regime (Jeon 2004, 26–33). In addition, it simultaneously attempts to expand the attributes of the military sector, which is almost the only systematized and efficient part of North Korea, to the other fields.

In order to carry out a socialist revolution, there must exist a leading

---

1) Refer to the New Year's Address by Kim Il-sung (January 1, 1994). To emphasize the U.S. responsibility on the nuclear issue, North Korea does not use the expression "North Korean nuclear issue" but "nuclear issue on the Korean Peninsula."

force, and the existing ideology has appointed the proletariat class to take on such a role. However, the military-first policy is a political measure that sees the military as the core of this revolution, since its power and spirit are considered superior to those of any other group in the state. The military is thought to be a successful materialization of the socialist collectivist nature not only due to its strong sense of revolution but also for being well organized and unified under strict discipline. This aspect is well reflected in the Party-military relationship. North Korea's stance is that "if the People's Army is strong it is possible to reconstruct the Party even if it were to fall apart, but if the Army is weak it is not possible to sustain the Party and furthermore, the whole country may collapse" (Eom and Yun 2006, 150–151). By indicating that the military is the foundation for the maintenance of both the Party and the state, there is an attempt to use its leading role as a means to solve the complex and difficult problems the country faces. In other words, by putting forward the military as a model to follow, the military-first policy aims to make other sectors of society unite in a similar way to exhibit a high degree of political capacity and thus ultimately attain system stability.

North Korea explains that military-first politics realizes the principle of "developing a military for a long period of time in order to use it for every aspect of national affairs (養兵天日, 用兵萬事)." Previously, it was "developing a military for a long period of time in order to use it for the most crucial moment (養兵天日, 用兵一時)." However, in the military-first era, the military is perceived as taking on the leading role not only in external security but in all sectors necessary for constructing a socialist paradise. By emphasizing the importance of the military in internal issues as well, the policy intends to control the military in an efficient manner and stabilize the regime. Such complex objectives reflect the reality that ever since the 1990s there have been limitations to attaining political stability solely through the North Korean people's voluntary support (Jeong 2009, 59–64).

Evaluation of Military Strategy in the Military-First Era

The military-first policy aims to achieve external and internal security, but despite these goals, its military strategy contains complicated structural drawbacks. North Korea's attempts to gain external security are instead further threatening it, for these efforts intensify the security dilemma around the Korean Peninsula. Similarly, its attempts to gain internal security obstruct the efficient allocation of resources and thus make the normalization of the economy, which is the key factor for regime stability, impossible. Such instability may ultimately result in the weakening of the regime and the state and present a challenge to power.

Security Dilemmas of Military Strategy in the Military-First Era

The military strategy of a military-first policy is making neighboring states feel threatened and react accordingly, thus worsening North Korea's security rather than improving it.[2] Although North Korea claims that its policy protects its autonomy and right to life, restrains war and promotes peace, and that it does not intend to undertake an arms race with other states or threaten them, it is producing the opposite result. Its military buildup is causing an arms race in Northeast Asia, which in turn becomes a boomerang that strikes North Korea. Consequently, Pyongyang has entered the vicious circle of security dilemmas and cannot help but continue to build up its hard power.

In addition, North Korea employs a peculiar reasoning that its production of nuclear weapons and missiles safeguards not only its own security but that of the Korean Peninsula and the entire Northeast Asia region from the American threat (Eom and Yun 2006, 164–165). Its stance is that without its military-first policy, war would have broken out multiple times, but due to nuclear weapons, the possibility of arms conflict on the peninsula has decreased and peace has been assured. North Korea also claims that due to the creation of a communal nuclear umbrella over Korea, its military-first policy" benefits the public and the security of South Korea (*Yonhap News* October 23,

---

2) Refer to Jervis (1978) on security dilemma.

2006). From North Korea's point of view, the Korean people's power against the United States derives from the North's possession of nuclear weapons, and, further, the North believes that these would be of great help even after unification (Eom and Yun 2006, 167). Such weapons, however, in fact place the Korean Peninsula at risk of nuclear war, make South Koreans feel increasingly threatened, destabilize the security order of Northeast Asia by increasing the probability of a nuclear chain reaction, and become a menace to the international nonproliferation regime. Caught in a vicious circle, these circumstances, in turn, all increase North Korea's insecurity.

In conclusion, the strategy of the military-first era is bringing about a security dilemma and preventing the state from resolving its problem of excessive military and security efforts. North Korea's military buildup and nuclear armaments foster an arms race of greater intensity surrounding the Korean Peninsula and also the American nuclear umbrella; therefore it eventually works as a great menace to the stability of the North Korean regime itself. While such a strategy may function to solidify North Korea's ego and maintain temporary security through the theory of balance of power, it does not help in dissolving the Cold War composition on the Korean Peninsula. On the contrary, it is intensifying competition and collision and becoming a huge obstacle for the signing of a peace treaty with the United States that Pyongyang so desires.

National Political Dilemma of Military Strategy in the Military-First Era

In the mid-1990s, after the Arduous March, the North Korean state system was not functioning adequately, which was an inevitable result of the military's insertion in national politics. It is true that the military took on a critical role while the economy as a whole suffered from food shortages and energy scarcity (Jeon 2004, 32). Kim Jong-il himself congratulated the People's Army, saying that its mobilization enabled the country to overcome the famine and open up the path for the construction of a strong and prosperous state (Jeon 2004, 168).

---

3) North Korea describes the strong and prosperous state as a military power, political ideology power, and cultural power (Kang 2002, 74).

But despite aiming for full control of the military and accelerating the construction of a strong and prosperous state by putting forward the military in both external and internal affairs, the military strategy of the military-first era hinders economic construction, the most crucial sector.[3] It is true that during a serious national catastrophe like the Arduous March, the mobilizing of the relatively well-organized and efficient military is helpful for bringing about social stability and reconstructing the impoverished economy. However, structural problems cannot be avoided once military strategy imposes the absolute priority of military logic in everyday politics and overpowers economic logic even in the construction of a strong and prosperous state. Such military strategy has become a burden to the state economy by causing excessive security concerns, and internally it has caused ineffective allocation of resources since they all are inordinately focused on the military. Instead of becoming a strong and prosperous state, North Korea is now a so-called wartime economy that impedes sound economic development.

This strategy's structural problems are revealed especially by the primary emphasis on the industrial sector of national defense in order to strengthen the military. Kim Jong-il claims that this focus is the lifeline for the construction of a strong and prosperous state, and that otherwise such an objective would be impossible to accomplish (Kang 2002, 70–71). North Korea is in discaccord with the traditional view regarding the military as a consumer of the economic sector, and states that peaceful conditions for economic development are only possible with a strong military. However, directing all units of economic construction for securing the factors of production for the national defense industrial sector distorts the distribution of state resources and hinders

Table1. North Korea's Economic Growth Rate, 1997–2008

Year	1997	1998	1999	2000	2001	2002
Economic growth rate (%)	−6.3	−1.1	6.2	1.3	3.7	1.2
Year	2003	2004	2005	2006	2007	2008
Economic growth rate (%)	1.8	2.2	3.8	−1.1	−2.3	3.7

■ Source: Bank of Korea

production in other fields. For Pyongyang, investments are allocated first to the defense industry, production from the military industry is given top priority in the production index and even in setting the national budget, and defense industry takes precedence. The military-first era's military strategy thus does not allow the economic structure to support an increase in factors of production like capital and labor; does not support improving production efficiency; and furthermore, does not provide space for new investment and production, since a smooth production-consumption-investment cycle is necessary for sound economic development.

Ultimately, this military strategy's emphasis on the industrial sector of national defense can be perceived as having political objectives rather than economic ones. This creates a dilemma that makes sound economic development impossible and hinders the construction of a strong and prosperous state; the logic of the military overpowers economic logic even in economic construction. Table 1 shows North Korea's internal dilemma through its low economic growth rates. Considering that a significant amount of capital was invested for about ten years after 1998 through South Korean aid and inter-Korean economic cooperation, it is possible to observe that the North Korean economy is facing a structural dilemma due to its military-first policy. Furthermore, there is a high possibility that it will face greater problems and more regime instability in the future.

Thus both in its external and in its internal aspects the military strategy of the military-first era has failed, since it has not been able to meet the goals that were originally set. Its main contents and structural problems can be summed up as follows.

Table2. Main Contents and Structural Problems of Military Strategy in the Military-First Era

Military-first Policy	Military Strategic Significance	Structural Problem
Political strategy placing the military as top priority in national affairs	Strategy for maintenance of external security	Aggravation of the security dilemma
Political strategy placing the military as core and main force	Strategy for internal system stability	Aggravation of national political dilemma

Gradual Decline and Seeking to Escape

Military strategy during the military-first era is facing dual problems of sustainability and durability. North Korea under its current ruling paradigm has two structural problems: it can neither give up its military-first strategy right away nor keep it indefinitely.

From the perspective of external military security, the military strategy of the military-first era aimed at the structure of a virtuous circle, so the state could build itself into a military power and achieve security for North Korea; however, in reality, a military buildup caused too much military growth and excessive security concerns, and ultimately an intensified security dilemma. As a result, North Korea entered into a long-term vicious circle, worsening its insecurity. From the perspective of domestic politics, the military strategy of the military-first era aimed for the virtuous circle that achieves an independent economy based on the organizational power and the efficiency of the army, and promotes national development and the pacification desired by the regime. Nevertheless, with economic deterioration, the domestic political dilemma has increased the possibility of regime crisis. Ultimately, military strategy in the military-first era unintentionally aggravated insecurity and weakened the basis of the regime rather than accelerating its efforts to develop into a strong and prosperous state, the original goal of Pyongyang.

Thus North Korea will not be able to resolve its deep-rooted problems and be assured of the survival of the regime and the state over the long term if it maintains its strategy of the current military-first era. In order to escape from its current foreign and domestic dilemmas, a correct decision by the Party's leadership is essential, and this should begin with the process of succession under way. The problems that North Korea has been facing can be classified into three: nuclear weapons, economy, and succession. The current problems of North Korea should be approached from a general perspective, not separately. The first step will be the new leadership's abandonment of nuclear weapons and strategic decisions that include changing from military-first politics to economy-first politics. It would be a most desirable solution for the current leadership of North Korea to make these strategic decisions on its own. We can imagine that Kim Jong-il might seek changes to some degree to lessen his son's burden, but it is almost impossible to expect him to do so in reality while he is in power today. Hence, such strategic decisions have to be accomplished in the process of succession.

Meanwhile, it has been difficult to assume that North Korea would make the strategic decision of abandoning its nuclear weapons program, considering its stance in the Six-Party Talks. Far from fundamentally resolving the overall North Korean nuclear issue, the process of the talks is inadequate. Making the decision to abandon its nuclear weapons and choose an economy-first strategy in the process of succession would be a desirable way for North Korea to proceed. However, even if the successor is not determined to give up nuclear weapons, North Korea will not be able to overcome the dilemma deepened by the military strategy of the military-first era. Unless North Korea stops repeating the cycle of cooperation and betrayal as it has used during the Six-Party Talks and the bilateral talks with the United States, the contradictions between its domestic and foreign policies will be aggravated, and the fundamental solution will be lost. Without the abandonment of its nuclear weapons, it will be difficult for Pyongyang to expect full-scale international aid, which is vital for its survival, and its crisis will be exacerbated. Taking a firm stand on military-first politics will do nothing desirable for North Korea but only hasten its decline.

Eventually, North Korea needs to make a strategic decision to give up its military strategy of the military-first era in order to resolve its problems in a fundamental way. The abandonment of the military strategy of the military-first era indicates the abandonment of nuclear weapons and the transition toward economy-first politics, which means resolving the dilemmas both in domestic politics and in foreign relations. Of course, immediate abandonment of the military strategy of the military-first era is neither realistic nor desirable. In consideration of the current situation of North Korea, immediate abandonment is likely to rapidly destabilize its domestic politics and weaken the perception of threat to the other states. Though the implementation would be conducted gradually, it is crucial for Pyongyang to abandon military-first politics and nuclear weapons and choose economy-first politics. For successful results, North Korea should strive to lessen military threats to the United States, South Korea, and the international community through negotiations on nuclear weapons and missiles, and at the same time, the international community also needs to make efforts to ease the sanctions previously imposed on North Korea. Also, it should work to establish a stable environment on the Korean Peninsula over the long term through the bilateral talks between North Korea and the other members of the Six-Party Talks, and the multilateral talks such as the Six-Party Talks, and make greater efforts to solve security dilemmas.

Policy Direction of North Korea

In preparation for the transformation of military-first politics into economy-first politics, North Korea has to drastically reduce its armaments and army and start putting efforts to resolve its excessive security concerns. It should hold disarmament talks with South Korea and, while maintaining its nuclear weapons, make a progressive effort at demilitarization that would transfer national resources concentrated in the army to the sector of the economy. Of course, such a strategic decision should begin with North Korea's declaration that it will abandon its nuclear weapons. Maintaining nuclear weapons will prevent North Korea from emerging from the strategy of the military-first era, and maintaining such military strategy will do the same for overmilitarization. In the end, under the situation where every sector of the state revolves around the army, it is impossible to establish a long-term national development strategy. The renouncement of the military strategy of the military-first era becomes a prerequisite for persuading the international community to help North Korea's development in the future. It is a new approach for North Korea to transform into a strong and prosperous state.

During such a confidence-building process, North Korea will be able to fulfill its promise to abandon its nuclear weapons through the Six-Party Talks and the bilateral talks. Since the existing Six-Party agreements have taken place without the full trust of North Korea, there has been a significant problem in implementation. Thus it is difficult to consider how North Korea's participation in the Six-Party Talks will directly lead to its strategic decision on the abandonment of its nuclear weapons. The effect of several agreements that the Six-Party Talks have made are limited to inducing North Korea's abandonment of its nuclear weapons. It is why the Six-Party Talks could stop the nuclear weapons program of North Korea but failed to make further progress in the basic stages of investigation and shutdown. Not building up trust with the other states, North Korea vowed to retain nuclear weapons, and the efforts of nuclear abandonment fell through.

Therefore, the leadership of North Korea has to decide to abandon the military strategy of the military-first era. The leaders will not be able to pursue a complete economy-first system during the initial escape stage, but the reform and open-door policy should be carried forward. It is a more realistic way of persuading North Korea to abandon its nuclear weapons strategically in the process of confidence building with other states and gradually carry out its decision rather than resolving it all at once. In reality, the complete abandonment of nuclear weapons should be attained in the next stage toward economy-first politics and reform. Thus the stage of escape, although the nuclear problem and economic problem will not be fundamentally resolved and some of the military strategy during the military-first era will still exist, becomes a basis for fulfilling the next full-scale implementation and reform since the North Korean government will have already made a strategic determination. Such movement will lighten the load of the new leadership of North Korea to a certain degree and help it to put promises like implementation and reform into action in the next phase.

In the stage of escape, full-scale disarmament specifically requires confidence-building measures in the military sector as a precondition. Building confidence between North and South Korea has been discussed for a long time but ended without substantive achievement. Article 3 of the October 4 Declaration from the Second Inter-Korean Summit Meeting in 2007 announced that the two Koreas will end their hostile relations, mitigate military tensions between them, and settle disputes through communication and negotiation. Furthermore, it specified the nonaggression duty, saying, "South and North Korea are opposed to any kind of war, and determined to firmly abide by the non-aggression duty." However, the situation that Pyongyang and Seoul do not trust each other devalued such a promise to an ostensible declaration. Meanwhile, North Korea insisted that the accomplishment of the North Korea–U.S. peace negotiation will naturally lead to an actual disarmament of the two Koreas, and that military confidence will be earned with the withdrawal of the U.S. forces in Korea (Ha 2000). However, examples from other regions, such as Europe, show military confidence building should take precedence in order to attain de facto disarmament (Han 2004). South Korea already made several suggestions in the UN Special Session on Disarmament in 1988; the confidence-building process included the first stage of building political confidence, the second of building military confidence, and the final goal of actual disarmament (Cho 2007). This confidence-building process should be fulfilled through mutual agreement between North and South Korea.

In this respect, with the abandonment of its nuclear weapons, North Korea's strategic decision should be transformed into a defensive military structure as part of its military strategy. Its military strategy is based on offensive strategies, such as total war, preemptive strike, mixed war, and swift strategy (Seo 2009). It is expected that North Korea will dominate a future situation by an early blitz attack and a mixed war of a regular and an irregular army and then expand to a war with powerful weaponry and mechanized units (Ministry of National Defense 2008). To put this plan into action, North Korea placed about 70 percent of its ground force south of Pyongyang and Wonsan and has developed offensive military structures to ambush enemy forces through overwhelming strength in the early phase of a war. Thus, in order to seek escape and fulfill de facto disarmament, North Korea should transform its offensive military strategy to a defensive one first. It also needs to relocate its offensive weapons system, start disarmament before the others do, and come up with measures regulating production and the import of offensive weapons.

The Roles of South Korea and the International Community

It will of course be hard for the North Korean government to trust promises that the international community will guarantee the security of its regime and provide grant aid in return for the abandonment of its nuclear weapons. From the realists' point of view, a country cannot trust another in the state of anarchy (Waltz 1979). This perspective explains the argument of North Korea that it cannot renounce its nuclear weapons because it does not believe the reassurances of the United States and South Korea. Thus, building mutual confidence with Pyongyang is the most important element in working toward a strategic decision. Until now, Washington and Seoul have framed their policies toward North Korea from the viewpoint of the hawks. But they need to understand the affairs surrounding the Korean Peninsula from the viewpoint of North Korea in order to dispel its worries about hostile U.S. policy and its perception of external threats in general. How to encourage North Korea to trust the international community also needs study from the viewpoint of North Korea. Both the Sunshine Policy and the policy promising North Korea's GDP of 3,000 U.S. dollars in return for the abandonment of nuclear weapons ended in failure. Both guaranteed full-scale aid and reform but could not be the fundamental solution to the North Korean issue because they did not have a practical plan to draw North Korea's strategic decision.

The international community, including the United States and South Korea, should guarantee the security of North Korea in the early phase of transformation from military-first politics into economy-first politics in order to build confidence. This guarantee assumes not only resolving external threats to North Korea but also maintaining North Korean domestic security during the leadership succession process. Signing the North Korea–U.S. peace treaty that North Korea is demanding in return for the abandonment of its nuclear weapons and the nonaggression pact by the United States and South Korea can be considered as ways to guarantee the security of North Korea. In addition, establishment of liaison offices could lay the groundwork for normalization of North Korea–U.S. and North Korea–Japan relations. Most important, establishing American liaison offices in North Korea has been discussed several times, the 1994 Geneva Agreed Framework being the most representative case. Improvements in North Korea–U.S. and North Korea–Japan relations are required to resolve problems related to excessive security and excessive military concerns. However, since North Korea cannot rely on U.S. extended deterrence as South Korea does, it should be able to reduce its perception of external menace by improving its relations with the United States. Such a measure might be insufficient to eliminate all Pyongyang's suspicions but can effectively help build mutual confidence.

China's active guarantee of the security of North Korea is one of the daring ways to resolve North Korea's perception of its external threats. After the early 1990s, North Korea's fear of external threats has rapidly increased since the former Soviet Union collapsed and Communist China lost its status as a security patron to North Korea in the process of systemic transformation. North Korea at that time lost the option of external balancing, namely, alliances, so it devoted all its strength to internal balancing, through armaments, in order to deal with the collapse of the balance of power on the Korean Peninsula.[4)] Thus it is important that a rising China actively secure the current North Korean regime under negotiations with the United States and South Korea. Considering Sino-U.S. or Sino–South Korea relations, China will hardly decide to provide extended deterrence to North Korea, but the improvement of Sino–North Korea security ties will be a great help in easing the fears that North Korea has, and also matches the strategic interests of China.

A Coevolution Strategy for the Advancement of the North Korean Military

Stage of Implementation and Reform

North Korea

The coevolution strategy of the North Korean military will begin when the regime decides to abandon its nuclear weapons first, passes the stage of escape, and enters the stage of transition and reform. Ultimately, the full shutdown of the nuclear weapons program should take place in the middle or end-

---

4) Refer to Waltz (1979, 168) and Morrow (1993, 208) on the relation between external balancing and internal balancing in balancing policy.

ing phase of transition and reform. In this stage, a new leader of North Korea would look like an enlightened dictator. He will ultimately concentrate on reform and an open-door policy for a market economy for the North Korean economy in order to resolve dilemmas in security and domestic politics. In this transition stage, the military-first and the economy-first governments will initially coexist. Thus, domestically, the powers of the military-first and the economy-first governments will form a balance of power, and the enlightened dictator should act as a balancer. Although neither the abandonment of nuclear weapons nor disarmament will have been completed yet, he has to fulfill them. The phase is important as the dilemma in security and domestic politics will only be resolved as national defense expenditures are reduced after renouncing military-first politics. In this stage, gradual transition and reform will be accomplished through a process of demilitarization with strategic changes in North Korea, and this transition will be completed at least by 2020, if not earlier.

North Korea should therefore undertake its disarmament in earnest. Drawing on military confidence built in the former stage, a detailed plan needs to be carried out. Most of all, before the Korean Peninsula becomes fully free from the conditions of the cease-fire, the mutual deterrence system needs to be maintained between the two Koreas while they still keep their military powers in balance toward the goal of disarmament. To put disarmament into practice, a mutual balance in military power should be maintained, with an application of the principle of mutual possession in equilibrium. It can be suggested that the one possessing more armed force reduce first to the level of the other, and then both sides start reducing to the same amount (Ha 2000). During this process, objections can be raised about the method of evaluating the levels of military power; however, such problems can be resolved by spot inspections and mutual observation by negotiation between Pyongyang and Seoul. North Korea has suggested that South Korea reduce both sides' troops to less than 100,000 and cut military facilities, but such a rapid disarmament is likely to exacerbate security conditions on the Korean Peninsula.

South Korea and International Community

Coevolutionary strategy emphasizes that such change should be encouraged not only in North Korea but also in other states, including South Korea and the United States, and finally should achieve harmony between the domestic and foreign situations of North Korea. If North Korea decides to abandon its nuclear weapons and military-first politics, the international community should welcome its decision and help it move toward a strategic transformation to economy-first politics. In the United States, there has been a suggestion and support for a "Grand Bargain," meaning that the international community should guarantee the security of the North Korean regime and provide full-scale economic aid to North Korea while North Korea in turn makes progress in abandoning its nuclear weapons and reducing its conventional weapons (O'Hanlon and Mochizuki 2003). Also, in the former stage, peace negotiations and a declaration of nonaggression between North Korea and the United States should be signed first. Then, the pledge can be implemented by the international community's support for the normalization of North Korea–U.S. and North Korea–Japan relations. Then there will be an expansion of full-scale economic support toward North Korea in the stage of transformation and reform. In the aftermath of the Cold War, North and South Koreas should have had serious disarmament negotiations and mutual disarmament as Europe did. These steps require the efforts of other states such as the United States, China, Japan, and Russia to continuously try to improve their relations with North Korea and create a condition in which the two Koreas can start mutual disarmament. Disarmament of conventional weapons and a cut in military strength on the Korean Peninsula will help to resolve most of the perception of inter-Korean military threat toward each other.

By all means, regular maintenance for and investment in the North Korean military is required in the process of transformation from military-first politics to economy-first politics since the transformation of national strategy has not been completed yet, and military-first politics and economy-first politics are still in balance. However, compared with former military-first politics, there will be more emphasis on the nonmilitary sector, where more national resources should be concentrated. Without efforts to resolve the problems of nuclear weapons and long-range missiles, and disarm conventional weapons and troops, the imbalance between military and economy will not improve. It is impossible for North Korea to make a new long-term national strategy without dissolving the excessive military emphasis of the military-first era. Consequently, the international community should support disarmament on the Korean Peninsula throughout the changing security environment and further guide North Korea in transforming from military-first politics into economy-first politics.

From the perspective of achieving peace on the Korean Peninsula and in East Asia, reform in the North Korean army and change in military relations between the two Koreas in the stage of transition and reform are destined to go through a transitional period. This circumstance reflects the distinctive military condition of the Korean Peninsula, which has been divided in tension for more than sixty years since the end of World War Ⅱ, and even after the Cold War, which ended in the 1980s. In consideration of the international order surrounding the Korean Peninsula, planning at a low level and gradual changes are more realistic than aiming for an idealistic peace. In such a transition stage, the international community should approach the reform of the North Korean military with an incomplete and passive approach, with more realistic conceptions of what peace will gradually look like, rather than applying a complete, rapid, and active plan for peace on the Korean Peninsula.

For example, in the stage of transition and reform, it is appropriate to plan for a new security condition on the Korean Peninsula from the viewpoint of "negative peace," rather than that of "positive peace" supported by Johan Galtung (Galtung 1996). Galtung has asserted that people should aim for a positive peace, in which elements of indirect and structural violence have militarily, politically, economically, and culturally disappeared and a mutual equilibrium is achieved, rather than a passive peace, which is the order that comes from the process of preventing a war and managing troubles. Kenneth Boulding classified peace into stable ones and unstable ones (Boulding 1979); the stable peace is the condition of peace that does not have the possibility of war and is without the need to prepare for a war, and the unstable peace has the possibility of war but is in a state of preventing a war through deterrence by power and regulation by negotiation. On the Korean Peninsula, positive and stable peace should be promoted in the long term, and the conditions ensuring permanent peace should also be aimed for. In the stage of transition and reform, the institutionalization process is the goal in order to manage the security dilemma. In other words, it is necessary to reduce the security dilemma in order to maintain current military strengths distributed around the Korean Peninsula, and ultimately to move to the institutionalization process of a security regime, preventing disputes and managing peace (Jervis 1983).

An international regime reflects the common expectation of agents on current issues in international relations.[5] A security regime is the process of institutionalization that states use to escape from security dilemmas in which no mutual confidence exists. For example, the Strategic Arms Limitation Talks (SALT) that the United States and the Soviet Union promoted was the means of mitigating the mutual arms race during the Cold War. In addition, the Nonproliferation Treaty can be evaluated as the nonproliferation security regime that effectively controlled the nuclear flow from the end of World War II to the 1960s. Coevolution strategy on the Korean Peninsula should be constructed as an institutionalization process for a transition period that ultimately makes progress in disarmament, reduces perception of threat, and resolves security dilemmas. Such formulation of a security regime is a relatively negative and unstable process, but it is more realistic in the militarily unstable security environment surrounding the Korean Peninsula.[6]

Nevertheless, the stage of transition and reform is just a period of transition, and a process toward ultimate transformation. It will be difficult for North Korea to become a normal state and survive in the international community should it keep staying in the stage of transition and reform in which not only does it make strategic decisions and pushes ahead with reform, and the military-first and the Sun-Kyung strategies coexist. For the survival and development of the state, Pyongyang must shed its military-first politics completely,

---

5) International regime is defined as "a set of explicit or implicit principles, norms, rules, and decision making procedures around which actor expectations converge in a given issue-area" (Krasner 1983, 2).

6) Refer to Jihwan Hwang (2009) on the necessity to institutionalize the peace regime on the Korean Peninsula and in Northeast Asia based on the concept of passive and unstable peace.

rule the state based on the rule of law and institutions, not by the dictatorship of the *Suryong,* or the Dear Leader, and substitute a planned economy with a market economy. In the military sector, the two Koreas should move beyond a negative and unstable peace through disarmament, settle into a positive and stable peace on the Korean Peninsula, and keep their military strength at a level that takes the security environment in Northeast Asia into consideration.

Transformation Stage

The last stage of the coevolution strategy includes a positive and stable peace, based on the full transformation of the North Korean military, on the Korean Peninsula and in Northeast Asia. This transformation stage is likely to happen between 2020 and 2030, and this period is the time when the domestic and foreign transformation of North Korea will be completed with full-scale changes in national strategy from military-first politics to economy-first politics.

Thus the institutionalization process of military security that has been promoted from the stage of transition and reform will be completed, and a stable peace will be settled on the Korean Peninsula. Such a development means not only the disappearance of the possibility of inter-Korean war but also a condition in which conflicts in Northeast Asia disappear and preparation for a war is not necessary. In addition, it indicates that the security environment on the Korean Peninsula will promote peace in Northeast Asia, and will be more secure in the process of developing peace in Northeast Asia. Such an image of the future suggests the concept of a security community in which the possibility of war among states almost vanishes in perception of an integrated community and adequate institutions and customs are established, as in Europe after the end of the Cold War (Adler and Barnett 1998; Deutsch 1968). For example, European states founded various conferences, including summit meetings and cabinet meetings, in order to deal with the changes in the international political environment and new security menaces. Through a series of institutionalization processes, leaders created the Organization for Security and Co-operation in Europe (OSCE) in 1995. In Northeast Asia, in the stage where multilateral security cooperation is institutionalized, security

Table3. Coevolution Strategy for North Korea' s Military

Stage	Escape from Crisis (First stage)	Transition and Reform (Second stage)	Transformation (Third stage)
Strategy of North Korea	• Declaration of abandoning military-first politics and nuclear weapons • Guaranteeing security of the regime, and the succession	• The coexistence of military-first and economy-first politics • Complete shutdown of nuclear weapons program • Efforts of the two Koreas for disarmament	• Defensive army • Security achieved by diplomacy
Measures of the International Community	• Promising the normalization of North Korea–U.S. and North Korea–Japan relations	• Supporting disarmament of the two Koreas • Trying to institutionalize security	• Completing disarmament by the two Koreas • Establishing a security institution

problems like building up military confidence and realizing disarmament could be stably resolved under the common recognition of the states.

In the stage of transition, North Korea will be a normal state that follows global standards. In such a situation, there would be almost no concern about security threats, so at this stage the measures implemented in the stage of transition and reform can be completed. Nuclear nonproliferation will be achieved by full-scale abandonment of nuclear weapons and long-range missiles, and finally denuclearization of the Korean Peninsula will be attained. North Korea, which will have completely resolved its tension with the United States, Japan, and South Korea, will participate in multilateral institutions in a positive way. Therefore, North Korea will enjoy a security environment that is free of the security insurance offered by China or the United States' nonaggression pact, and implementation of disarmament by the two Koreas lets both of them keep their military at minimum. Economically developing North Korea would maintain its necessary army at a minimum, but its power would be limited to defense and not expected to threaten security on the Korean Peninsula and in Northeast Asia.

Hence, security guarantees on North Korea will not depend on military strength but on diplomacy and Northeast Asian multilateral security institutions. Of course, the Northeast Asian security environment and multilateral institutions can be fluid, relying on Sino-U.S. or Sino-Japan relations. However, it is certain that peace in Northeast Asia will be promoted when the North Korean issue does not threaten the Korean Peninsula and the security of Northeast Asia. ■

■ References

Adler, Emanuel, and Michael Barnett. 1998. *Security Communities*. Cambridge: Cambridge University Press.

Boulding, Kenneth. 1979. *Unstable Peace*. Austin: University of Texas Press.

Cho, Seong-Ryoul. 2007. *Han-ban-do pyeonghwa-che-je* [Peace Regime on the Korean Peninsula]. Seoul: Pureunnamu.

Chun, Chaesung. 2009. "Moving from a North Korean Nuclear Problem to the Problem of North Korea." EAI Issue Briefing No. MASI 2009-03, June 1.

Deutsch, Karl W. 1968. *The Analysis of International Relations*. Englewood Cliffs, NJ: Prentice Hall.

Eom, Gukhyeon, and Geumcheol Yun. 2006. *Jo-seon-ban-do pyeonghwa-bo-jangmun-je* [Issue of Peace Keeping on the Korean Peninsula]. Pyongyang: Pyongyang Publication.

Galtung, Johan. 1996. *Peace by Peaceful Means*. London: Sage Publications.

Ha, Young-Sun. 2000. "Han-ban-do-ui pyeonghwa gu-sang" [Peace Design on the Korean Peninsula]. In *Bun-dan-eul neom-eo tong-il-eul hyanghae: nambug-hae-oe-hag-ja tong-il-hoe-ui 5nyeon-ja-lyo-jib* [Toward Unification: Five years of Meetings for Unification by Scholars from the Two Koreas]. Young-Chul Paik, ed. Seoul: Konkuk University Press.

Han, Yong-Sup. 2004. *Han-ban-do pyeonghwa-wa gun-bi-tongje* [Peace on the Korean Peninsula and Arms Control]. Seoul: Bagyeongsa.

Hwang, Jihwan. 2009. "Han-ban-do-pyeonghwa-che-je gu-sang-ui i-sanggwa hyeon-sil" [Ideal and Reality of Designing Peace Regime on the Korean Peninsula]. *Peace Studies* 17, 1.

Jeon, Deokseong. 2004. *Seon-gun-jeongchi-e dae-han li-hae* [Understanding of Military-First Politics]. Pyongyang: Pyongyang Publication.

Jeong, Seongjang. 2009. "Gimjeong-il jeonggwon-ui seon-gun-jeongchi-wa gwon-lyeog-el-li-teu" [Military-first Politics of Kim Jong-il Regime and Power Elites]. *Gmjeong-il-ui bug-han, eo-di-lo ga-neun-ga* [Kim Jong-il's North Korea and Where it is Going]. Association for Contemporary North Korean Studies, ed. Seoul: Hanul Academy.

Jervis, Robert. 1978. "Cooperation Under Security Dilemma." *World Politics* 30, 2.

————. 1983. "Security Regimes." In *International Regimes*, Stephen D. Krasner ed. Ithaca, NY: Cornell University Press.

Kang, Sung-Gil. 2002. *Seon-gun-si-dae-ui jo-gug-eul ga-da* [Going to the Homeland of the Military-first Era]. Pyongyang: Pyongyang Publication.

Krasner, Stephen D., ed. 1983. *International Regimes*. Ithaca, NY: Cornell University Press.

Ministry of National Defense, Republic of Korea. 2008. *National Defense White Paper.* Seoul: Ministry of National Defense, Republic of Korea.

Morrow, James D. 1993. "Arms versus Allies: Trade-offs in the Search for Security." *International Organization* 47, 2: 208.

O'Hanlon, Michael, and Mike Mochizuki. 2003. *Crisis on the Korean Peninsula: How to Deal with a Nuclear North Korea.* New York: McGraw-Hill.

Seo, Joo-seok. 2009. "Bug-han-ui gun-sa-je-do-wa jeongchaeg" [Military System and Policy of North Korea]. *Bug-han-che-je-ui i-hae : je-do-wa jeongchaeg-ui ji-sog-gwa byeon-hwa* [Understanding North Korean Regime: Continuations and Changes in the System and Policy]. Seoul: Myung In Publishers.

Waltz, Kenneth N. 1979. *Theory of International Politics*. Reading, MA: Addison-Wesley.

Woo, Seongji. 2007. "Gimjeong-il-si-dae bug-han-ui gug-je-gwan-gye-lon i-hae-leul wi-han si-lon" [Commentary for the Understanding North Korea's International Relations Theory in Kim Jong-il Era]. *Korean Journal of International Studies* 47, 4.

CHAPTER 5

A Strategy for the Advancement of the North Korean Economy

Dongho Jo

Introduction

THE KIM JONG-IL ERA OF NORTH KOREA IS AN ERA OF MILITARY-FIRST politics. Military-first politics is a survival strategy of North Korea that directs every policy area. It was chosen while North Korea went through extreme difficulties derived from the collapse of the socialist bloc in the early 1990s and the Arduous March in the mid-1990s. No matter how dire the situation was, Kim Jong-il could not possibly abandon socialism and introduce an active opening and reform policy. Thus a military-first strategy was an inevitable choice for North Korea in the 1990s.

Necessarily this military-first strategy brought about changes in traditional economic strategy. In the Kim Il-sung era, economic strategy had made developing heavy industry and simultaneously developing light industry and agriculture a top priority. According to the military-first principle, however, the focus shifted from heavy industry to the military industry. The principle also implied that all economic fields had to be sacrificed save the development of the military.

What were the reasons for adopting the economic strategy of the military-first era and what were the underlying motives? Can such an economic strategy succeed? If it is bound to fail, what kinds of strategies could come next?

What would be the role of South Korea and neighboring countries to expedite the economic development of North Korea?

This chapter seeks to answer these questions. It concludes that North Korea's economic strategy in the military-first era was positive for handling economic difficulties in the short run but could not help producing negative effects on the economy in the long run and finally bringing about the collapse of the North Korean regime. Therefore, this paper suggests that North Korea must abandon its existing economic strategy and adopt a new one if it wants to avoid collapse in the future.

Economic Strategy in the Military-First Era

Concept and Contents

According to North Korea, the military-first era is defined as "the era where all problems occurred in the process of revolution and reconstruction should be solved by the military-first principle" (Kim 2005, 58). Giving priority to the military means that "military affairs take precedence over others and strengthening the military industry should be treated preferentially" (Kang 2002, 19). North Korea has been propagating the military-first principle as a "new and ingenious idea" created by Kim Jong-il. In this sense, North Korea is criticizing Karl Marx by implying that he overlooked the importance of the military industry. North Korea is also criticizing Vladimir Lenin and Joseph Stalin in their emphasis on heavy industry and lack of "the right view on the strategy of the economic development" (Lee 2003, 8; H.S. Kim 2004).

It was natural that North Korea's economic strategy had started to focus on the military industry as the military-first era emerged. Such a strategy is called the economic strategy in the military-first era, which means putting the top priority on developing the heavy industry and simultaneously developing light industry and agriculture. According to the *Gyungje Yeongu* or *Economic Studies*, the most important North Korean quarterly economic journal, Jae-Seo Kim (1999) is the first article that emphasized the importance of the military industry and Seong-Hyeok Lee (2001) is the first one that connects the military-first principle and the military industry. The first article that includes "military-first" in the title is Dong-Nam Kim (2001), but it did not actually cover the military industry. The review of the articles in *Economic Studies* concludes that the discussions systematically analyzing the reciprocal relationship between the military-first principle and the military industry had not been initiated until the work of Lee (2003) and Park (2003).[1]

Although economic strategy in the military-first era purports to mean that light industry and agriculture are to be developed simultaneously, what it actually means is that light industry and agriculture should be sacrificed in order to develop the military industry.[2] In other words, the military industry is the "lifeline of building a strong and prosperous country" and it is the "fundamental economic principle in the military-first era" to develop the military industry. Thus "the government expenditure must be invested in the military industry before any other industry" (J. S. Kim 2004, 13–14) and "the necessary inputs such as workers, facilities, raw materials, and electricity must be provided preferentially for the military industry" (Lee 2003, 5). Furthermore, Cho (2005) explains that what economic strategy in the military-first era really means is that the advancement of light industry and agriculture is needed to reinforce the military industry and that the military sector is the top priority among all sectors of the economy.

North Korea is using the term "military industry" as the "munitions industry." For instance, Kim (2005, 109) defines "the military industry" as an industry to supply modern arms. In fact, North Korea divides its production sectors into three categories: munitions production, producer goods production, and consumer goods production, and the munitions industry is defined as the industry that produces materials used for military purposes. Thus North Korea says that "the munitions industry is the same as the military industry

1) According to the recent study, Kim Jong-il presented a guideline of "developing the military industry first and developing the light industry and agriculture simultaneously" on Sept. 5, 2002, and after the process of theorization, it was officially announced as a remark titled "let's accomplish thoroughly the economic strategy in the military-first era proposed by the party" on Aug. 28, 2003 (Han 2009, 149–150). Kwon (2008) also analyzes that the theorization process of the economic strategy in the military-first era started after Sept. 2002.

2) "In the military-first era, the military affairs should come first, the military industry should be treated first, and developing the military industry is the most vital task" (Kim 2005, 109).

since the munitions industry contributes to defending North Korea from military attacks by the imperialist countries" (H. S. Kim 2004, 2).

Logic

What is the logic behind the economic strategy in the military-first era and how does North Korea justify such a strategy? According to North Korea, the goal of economic activities in a socialist society is to guarantee the independent and creative lives of workers just as it is to produce surplus value for the properties of the bourgeoisie in a capitalist society. The independent and creative lives of workers consist of two parts. First, people have to be independent as the owners of the society and their independence is secured by protecting socialism from the invasion of imperialism. Second, people must enjoy abundant and civilized lives that are guaranteed by fulfilling material needs.

North Korea explains that the first part is more important than the second part because it is impossible to improve people's living standards unless socialism can be protected from the aggression of imperialism. In other words, people's wealth is meaningless if they remain slavish. So any other industry cannot be more important than the military industry and without its development, national security cannot be ensured and economic improvement cannot be achieved. It is to say, "we can live without a lollipop but not without a bullet" (D. H. Kim 2004, 6) and "it is human nature to prefer living in an independent country even if food and clothes are much worse than living in a colonized country."

Undoubtedly light industry and agriculture will decline if the military industry is regarded as the top priority, since the funds to invest are limited. As a result, they will influence people's living standards in a negative way. But North Korea is insisting that "the military-first principle is not sacrificing people's living standards by focusing on the military industry. It is not only to protect the nation and socialism but also to provide better lives for the people (Park 2004, 16). However, North Korea does not present the specific logic of how it is going to improve people's living standards.

Then how will economic strategy succeed in the military-first era? North

Korea claims that the voluntary work of workers based on their revolutionary ideas is the basic means and that "revolutionary military spirit" makes them achieve a higher level of effort for their voluntary work (Park 2003, 8). To answer the question why "revolutionary military spirit"is required, North Korea explains that every era needs the spirit of the times and thus they need a revolutionary military spirit as North Korea enters the new era of the military-first principle (Cho 2003, 5).

Economic Strategy in the Kim Il-sung Era

Economic strategy in the Kim Il-sung era can be summarized as putting top priority on developing heavy industry and simultaneously developing light industry and agriculture. So the difference of the economic strategies between the Kim Il-sung era and the military-first era is whether top priority is placed on heavy or military industry.

North Korean literature indicates that the economic strategy of the military-first era is not separate from that of the Kim Il-sung era and in fact inherits the latter. The reasons can be explained as follows. First, the military industry is closely related to heavy industry because heavy industry is the foundation of the military industry, and the military industry induces the rapid development of heavy industry. Second, the military and heavy industries share common factors in terms of physical and technological characteristics. For instance, the weapons industry, which is the core of the military industry, is the machine industry, which belongs to heavy industry. Third, the development of heavy industry is a precondition for the development of the military industry since the products of the heavy industry, such as fuel, electricity, equipment and facilities, are the inputs of the military industry.

In addition, North Korea claims that Kim Il-sung did actually stress the importance of the military industry and endeavored to reinforce it. For example, he noted that "the modern military industry is one of the most important areas in building self-reliant national economic foundation" and he accentuated the "importance of the military industry in his entire leadership period" (Kang 2004, 2). There also exist articles enumerating the cases of building factories for the military industry since 1945 (Sohn 2005). But the economic strategy in the military-first era is different from that of the Kim Il-sung era since it is clear that the top priority in each strategy is not the same. The reason why North Korea asserts two strategies is basically identical, based on the realistic environment in which they cannot deny the legacy of Kim Il-sung.

Evaluation of Economic Strategy in the Military-First Era

As mentioned above, the economic strategy in the military-first era was a product of North Korea's serious difficulties in the 1990s, both internal and external. The upsurge of crisis and tension from domestic and foreign sources caused the advent of the military-first strategy, as North Korea has conceded.

In the mid-1990s, the concern North Korea had about the collapse of its regime reached its climax. Kim Il-sung died suddenly, the so-called Arduous March began, the socialist bloc had fallen, and the United States became a superpower. In addition, China, which had been in a tight alliance with North Korea, accelerated its efforts to improve its relationship with South Korea. Thus North Korea had no one to depend on. In this situation North Korea concluded that military power was the only answer for defending its regime by itself. This idea was realized in military-first politics.

Still, it was not the right choice and North Korea should have adopted an economy-first, not military-first, politics. As long as North Korea focuses on its military industry, its economic strategy is destined to fail. Furthermore, the economic strategy in the military-first era is nothing but a slogan and it does not mention any specific ways to proceed with the strategy.

In fact, the way to promote economic growth is the same in both a socialist planned economy and a capitalist market economy in the sense that economic growth requires inputs such as labor and capital or technical progress. But since North Korea is in a state of full employment, it is impossible to increase the workforce. The poor economic condition as well as the closed-door economic policy also limits technical progress. Thus, how to acquire capital is the key to economic growth and the needed economic strategy should provide a blueprint for inducing foreign capital.

But economic strategy in the military-first era has not mentioned ways to mobilize capital domestically and internationally. And the military sector cannot invite foreign investments by its nature. A limited budget, furthermore, focusing on the military sector should be followed by the shrinking of the other sectors in the economy.

Instead, the economic strategy in the military-first era suggests several other measures such as science and technology, trade, and production efficiency. But none of these are likely to bring about positive outcomes under North Korea's present economic situation. For instance, the science and technology North Korea has been emphasizing is understood in practice as the modernization of facilities (Yang 2002, 19). However, it is extremely difficult due to the lack of capital. Increasing trade volume enough to contribute to the economic growth is also almost impossible unless North Korea carries out active opening and reform programs. At last, North Korea introduced "the July 1st economic measures" in 2002 and permitted the establishment of "comprehensive markets" in 2003.

Some studies, such as those of Kwon (2008) and Han (2009), apprehend that the economic strategy in the military-first era and "the July 1st economic measures" were carried out almost simultaneously as they noticed the timing that the economic strategy became known to the public. For example, Han says that "Kim Jong-il actually planned to relate the prioritizing [of] the military industry to the improvement of economic management in the civilian industries, and thus his plan was a mixture of reformative and conservative policies" (Han 2009, 155). But it seems unreasonable to think that one person would plan two policies that are contrary to each other at the same time. Therefore, it would be more likely that the economic strategy in the military-first era was initiated before "the July 1st economic measures," although its official announcement was delayed. In fact, Kim Jong-il ordered at the conference with officials of the central committee of the Party on January 1, 1997, that "even the economic projects should be related to strengthening the military" and also stressed at the same occasion on January 1, 1999, that "the economically strong country cannot be built without the military industry. Thus, it should be given top priority" (Lim 2009, 26–27).

If we interpret that the economic strategy in the military-first era and the "July 1st economic measures" proceeded not simultaneously but consecutively, it is much easier to understand the intention of the economic policies of North Korea. To Kim Jong-il, the military-first politics cannot be discarded under any circumstances, because they represent the slogan of his era. As a result, the economic strategy in the military-first era cannot also be retracted. Thus the "July 1st economic measures" were executed to improve economic management by partially introducing several market-based functions within the economic strategy in the military-first era. In this sense, we can conclude that the "July 1st economic measures" were not to transform the North Korean planned economic system to the market-oriented one but to complement shortcomings of its system. Such measures must be needed in order to build an economically strong country, because an economic strategy focusing only on the military industry cannot achieve the goal. Therefore, the "July 1st economic measures" and the establishment of the "comprehensive markets" were to guarantee the success of the economic strategy in the military-first era, and they have to be analyzed as the introduction of reformative policies for complementing conservative ones, not the mixture of reformative and conservative policies.

The "July 1st economic measures" and "comprehensive markets" did bring about positive influences on the North Korean economy, since they provided firms and workers incentives to produce more and opportunities to be involved in economic activities through the markets. In fact, the setup of markets entails the creation of additional value. For instance, whereas in the past there was nothing more when the government produced and distributed 100 dollars worth of wheat, now people do not just simply consume the wheat but they produce and sell bread or noodles with that wheat at the price of 120 dollars in the market. Consequently, 20 dollars worth of value-added have been created, and the GDP and the economic growth rate rise that much.

However, they also resulted in negative effects. People came to prefer business in the markets over working in their firms because they could not be paid properly and regularly at their workplaces due to the reduction of operations caused by the insufficient supply of electricity and raw materials. Thus the attendance rate went down to 10 percent, the socialist education along with debate about work attitudes that used to take place every day was now held only once a month, and it became common knowledge that about 90 percent of workers performed their economic activities in the markets (Park 2008, 7–8). Even a class of merchants appeared and a spurt of industrial capital emerged (Kim 2008). Furthermore, various forms of illegal and private economic activities were on the rise. It is estimated that about 20 percent to 30 percent of agricultural production is produced on privately cultivated land, and that a portion of the production of firms for market sales takes up about 20 percent to 30 percent for consumer goods and about 5 percent for producer goods. In addition, illegal economic activities such as private financing, smuggling, trading foreign currencies, real estate transactions, and private corporations are rising.

Although what North Korea intended by introducing the "July 1st economic measures" and "comprehensive markets" was to normalize and strengthen the socialist planned economic system, the negative effects such as expansion of market activities became too big to ignore and made the North Korean authorities worried about the possible collapse of the regime. As a result, North Korea started to control market activities and to withdraw the "July 1st economic measures." For instance, North Korea tried to decrease market size by reviving the distribution system in 2005 and banned sales in the market by adult males over the age of 17 in 2006. In 2007, the age allowed to sell in the market was limited to over 40 years old, and several other policies were activated such as liquidation of unofficial trading companies, control of sales items and prices in the market, and prohibition of sales outside the marketplaces. Finally an announcement was made in 2008 that the market would be closed down from 2009 on.

However, reality could not change as the North Korean authority intended, because people who had already experienced the efficiency of the market system and could no longer live without the market raged against these restrictions on a serious scale. Even if there were no reaction by the people, the North Korean government could not have arbitrarily coerced the close-down of markets because it would only have led to economic chaos and hardship as long as the North Korean government was incapable of restarting distribution.

After all, economic strategy in the military-first era, which mainly focused on political survival rather than on economic growth, was doomed to failure. From the beginning, in particular, it had a fatal limitation in the sense that it did not give answers to how to induce foreign capital that was the core for economic growth. As a result, since the introduction of the economic strategy in the military-first era, the North Korean economy has confronted a situation where legal and illegal activities exist together and where planned and market systems coexist as well. It is this situation that upholds the living of North Korean people and positively influences the North Korean economy. But it also means the start of the collapse of the socialist planned economic system. Thus we can conclude that the current situation of the North Korean economy is in the process of growing and collapsing simultaneously.

Inevitable Decay and Groping for Escape

Starting around 2010, North Korea has shown a series of changes in economic strategy in the military-first era. For instance, the New Year's Joint Editorial of North Korean newspapers in 2010 emphasized to the highest degree the importance of light industry and people's living standards. This was very unusual. However, such a stress on light industry has continued even today.

It seems that there have been debates and conflicts among the ruling elites in North Korea concerning the economic strategy in the military-first era as it failed to achieve the goals originally anticipated. Such a symptom can be found in the articles of *Economic Studies*. For instance, we can see the difference in policy priorities between the military industry and heavy industry even if it is slight so far. The original position is that heavy industry should serve a military industry as directed by the economic strategy in the military-first era. One of those examples is "it is the fundamental principle that every industry must provide any materials in any condition if they are required in the military industry" (Hwang 2006, 14). However, the implication is different in the positions such as "the military and civilian industries are closely related and affect each other" (Baek 2006, 16–17) and "[a] bigger burden is laid on the heavy industry strategy in the military-first era" (Cho 2007, 11). In addition, some articles emphasize less the importance of the military industry than the traditional ones. Those examples are that "the heavy industry should serve equally for the light industry, agriculture, and the military industry" (Oh 2008, 11–14) and that an "economic strategy in the military-first era means an exact balance among the light industry, agriculture, and the military industry" (Lee 2008, 19).

It is not clear that North Korea has really decided to change an economic strategy that focuses solely on the military industry. All we can say for sure is that North Korea has begun to emphasize the importance of light industry and people's living standards. It would be mainly because the year 2012 is coming, the year by which Kim Jong-il himself has declared North Korea will have built a strong and prosperous country. North Korea has insisted that it is competing by building an economically strong country because North Korea is already strong in military, politics, and ideology.

It implies that North Korea has now every incentive to focus on improving economic conditions. First, the North Korean people may not be persuaded by the announcement that they have become a strong and prosperous country without an improvement in their living standards that they can really feel in their daily lives. Second, building the strong and prosperous country is the last task of Kim Jong-il to finish his era magnificently. He will be remembered in history as the "great leader who has built the strong and prosperous country" as his father is remembered as "great leader who built the self-reliant country." Third, Kim Jong-un himself also has an incentive to focus on the economy. Although he was appointed as a successor by his father, he is not well known to the people and thus has to show them his ability as a potential leader. Furthermore, helping with his father's last task will contribute to solidifying the power succession from his father.

In order to construct a solid foundation for economic growth, however, the fundamental change of the economic system and policies is an absolute prerequisite. The history of China and the Eastern European countries suggests that only active reform of the system and wide opening of the economy can bring about a self-sustaining foundation for economic growth. Nonetheless, in the era of Kim Jong-il such a change cannot be anticipated.

As a result, it seems that North Korea will have to rely only on the efforts to induce foreign supports and investment in order to achieve its goal in 2012. The recent announcement of raising the status of Naseon City and the establishment of Chosun Daepoong International Investment Group probably are parts of such efforts. Nevertheless, in a situation where the nuclear issue is not solved, it would be impossible to invite foreign assistance as much as North Korea wants. Moreover, foreign firms will not launch big investment programs if North Korea does not initiate economic reform and liberalization. Even if some amount of investment is made, the effects of the induction of foreign capital will hardly be seen as long as North Korea sticks to its current economic strategy.

Therefore, the North Korean economy cannot but help decline if military-first politics remain unchanged. Consequently, the abdication of military-first politics, the reform of the economic system, and the liberalization of the economy are the only ways to prevent the possible collapse of the regime and to develop the economy. Of course, such a change will be expected in the post–Kim Jong-il era since he cannot abandon the military-first politics that is the slogan of his era.

The Strategy of Coevolution

Transition and Reform Stage

North Korea

In order to keep North Korea alive, Kim Jong-un or the next North Korean leadership has no other alternative but to make the decision to give up nuclear programs and military-first politics. In return, he will receive economic support and investment from the international community, including South

Korea, and these funds and goods will function as a driving force for rapid economic growth of North Korea.

However, an inflow of foreign capital alone cannot facilitate preparing a self-sustaining foundation for economic growth unless an environment that can effectively utilize the capital is fully furnished. Consequently, the next leadership needs to promote reform of the economic system as a new means of economic survival. On a contextual level this means North Korea has to initiate the transition from its traditional socialist planned economic system to the capitalist market-oriented one, and on an ideological level this means that military-first politics should be changed to economy-first or civilian-first politics.

How to proceed with the transition of the economic system is usually divided into a gradual approach and a big-bang approach, or shock therapy. For North Korea, a gradual approach is seen to be a better one. First, in the transitional period, serious side effects and chaos are inevitable and the risks are much greater in the case of a big-bang approach. North Korea, which has been suffering economic difficulties during the last twenty years, is not likely to endure those negative impacts. Second, North Korea has never experienced the capitalist market system. Thus, the radical change of the economic system would be too risky and lead to the collapse of the state itself. Third, there exists South Korea as an economic alternative to the North Korean people. An extensive and full-scale transition would make the North Korean people see no difference between the two Koreas economically and start to ask for unification. Thus, the North Korean leadership would like to avoid such a situation.

Even in a gradual approach, the core of the transition is a change from the state-owned property system to the private property system and the substitution of a planned economic system with an autonomous market mechanism. Therefore, the transition of the economic system fundamentally differs from the economic reforms that have been exercised in the former socialist countries in Eastern Europe, because they were basically improvements within the system rather than reforms of the system and therefore can be characterized as "system-preserving reforms." And all of these reform efforts turned out to be a failure.

In this context, the privatization of property rights and the introduction of market mechanisms is an assignment absolutely crucial in the future reform of North Korea. This means that the reform policies the Eastern European countries adopted as they initiated their economic transition in the early 1990s should also be taken by the North Korea. The examples of those reform policies are privatization, price liberalization, financial reforms, trade liberalization, macroeconomic stabilization, and building social security networks. In the case of North Korea, additional reform policies such as annulment of the North Korean self-reliant economic management, joining international financial institutions, and adjusting industrial structures should also be promoted. However, as already pointed out above, simultaneously promoting all of these reform policies is neither reasonable nor realistic. Therefore, it is required to set priorities among the reform policies to be promoted.

In the case of privatization, huge side effects will occur if all stateowned properties are handed to the private sector at once. The North Korean people do not have the economic ability to buy those properties and foreign firms that can purchase a tremendous amount of the North Korean properties could not easily be found. Even if there exist a number of foreign firms that have the will and ability to buy North Korean assets, it is very likely that the North Korean people, familiar with the socialist system for decades, will have negative feelings concerning the reform process as their self-esteem feels damaged due to the leakage of national wealth and the predominance of foreign capital. On the other hand, a free distribution to the North Korean people can hardly be thought of as a reasonable alternative considering the huge financial resources required for promoting reform. Consequently, in the first stage, it would be rational to start with small-scale privatization. For instance, private rights should be acknowledged in small and medium-sized manufacturing firms and service industries such as restaurants, snack bars, stores, and gas stations. Along with the freeing of entrance and exit, the introduction of a law of fair competition is necessary for the smooth promotion of such a privatization.

Price liberalization is also a core part of an economic transition. However, full-scale price liberalization that would nullify government control on all prices at once will bring about an excessive shock to the economy because no economic players in North Korea have experienced a situation in which prices are freely determined in the market without intervention of the government. Moreover, serious inflation will be an inevitable result of full-scale price liberalization and thus the North Korean people will be in great reaction against the economic transition. Therefore, it would be better to liberalize prices gradually, reflecting the situation of the economy. Items such as daily necessities and strategic goods should be the last ones in the price liberalization process. Meanwhile, it is important that the North Korean government control the rent-seeking opportunities that are caused due to the dual structure of prices.

The reform of the financial sector is very important in producing and circulating the capital required in the real sector and achieving macroeconomic stability by enabling the absorption and control of excessive liquidity that is widespread within the North Korean economy. The most urgent task in the financial reform is the abrogation of the mono-bank system as seen in the transition period of the Eastern European countries. Although other financial reform plans can be executed on a gradual basis, it is strongly advised to complete the establishment of the two-tier banking system in the very beginning stage. Considering the financial institutions of North Korea today, it would be reasonable to transfer the functions concerning trade and foreign exchanges from North Korea's Central Bank to the Trade Bank and to establish new commercial banks to take over the commercial functions of the Central Bank.

The normalization of currency functions is also necessary. Traditionally in North Korea, currency has been perceived only as a simple means to execute state plans. However, currency should restore its original functions such as a means of trade, a medium of exchange, a store of wealth, and a measure of value. For this reason, cash transactions should be institutionalized for all cases and the traditional "no-cash-transaction" principle should be abolished.

Moreover, the problem of insolvent debt, which has also been a fruit of the old regime, should be solved quickly in order for the North Korean banks to perform their functions normally. As the experiences of the Eastern European countries clearly show, the whole banking sector can be paralyzed unless such a problem is solved in the beginning stage. Hence, it is advised to establish a special institution exclusively responsible for solving the problem of insolvent debt in a timely manner. By doing so, the North Korea banks should be given chances to survive from competition with the South Korean and foreign banks. Meanwhile, since the unsolved liabilities occurred in the 1970s from the loans from the Western banks are characterized as national liabilities, it is fair that the Central Bank take them over.

Improvements in exchange rate policies are also needed. The exchange rates of North Korea are divided between the official and the unofficial ones, and the reality is that the gap between the two is very large. Unifying them into one official exchange rate is necessary. The level of exchange rates should be able to reflect both domestic and global economic situations. Between the fixed and flexible exchange rate systems, the latter is better in the beginning stage of the transition, considering that macroeconomic stabilization is one of the top priorities.

At the core of the reform of the public finance sector lies abolishing the soft budget constraint. Since one of the most significant goals of the economic transition is to improve efficiency, it is a precondition to alter the soft budget constraint into the hard budget constraint. However, it is reasonable to put in action the hard budget constraint gradually, starting with the small and medium-sized firms. If the hard budget constraint is applied to all the firms at once, the economy is very likely to be thrown into serious difficulties due to widespread bankruptcy. It is also important to integrate the budgets belonging to the military and the Party into the budget of the cabinet. Because it is estimated that the military and the Party possess a large amount of available budgets, it is impossible to either promote reform or strive for economic development when their budgets are left alone. Furthermore, if those budgets are not absorbed, the military and the Party can rise as a seriously challenging threat to reforms based on economic power when they believe that their interests are being trespassed.

The reform of the trade sector should be pursued as a part of the development strategy for the underdeveloped economy. First of all, any firms that are interested in international trade must be allowed to partake in order to activate international trade. In addition, the nontariff barriers such as import approval and quotas should be removed. Higher tariffs may be imposed on consumer goods in the beginning period in order to stabilize the economy. Import regulations may also be necessary to protect some strategic industries even if they are inefficient. In that case, however, it is desirable to announce clearly from the beginning that such a protection policy is temporary to secure their competitiveness.

Stabilization of the macroeconomic condition is fundamental to promote transition and reform. It should be remembered that the Eastern European countries like East Germany, the Czech Republic, Slovakia, Hungary, and Poland, which achieved macroeconomic stabilization in the early period, were able to reconstruct their economies relatively smoothly. But countries like Russia, Romania, and Bulgaria failed to control inflation and suffered economic woes as a result. Strengthening the social safety net is also necessary for securing supports for the economic transition. Because the social safety net in North Korea has been weakened considerably since the Arduous March and the transition itself from the socialist to the capitalist regime will produce an additional burden on people, the minimum level of social welfare and security must be guaranteed.

The reform of the economic ideologies peculiar to North Korea must be carried out in the beginning stage. In industrial management, the Daean management system should be abolished and changed officially to a manager-centered system. In agricultural management, a contract system based on family units must be introduced to improve productivity, although the collective farm system is basically maintained in the beginning period. Such economic slogans as "self-reliance" and "self-support" should be substituted for rational ones like "comparative advantage" and "competitiveness." In addition, more stress must be put on material incentives than on moral ones.

The reformation of the industrial structure must be done in the beginning stage. A gradual reformation according to a well-designed schedule starting with the industries that have the least impact on the economy is more desirable than a full-scale and simultaneous one. The most urgent task in this transition and reform stage is to reduce heavy industry, including the military, because it has taken too large a portion in the industrial structure, and its inefficiency due to outdated facilities and low productivity is now too serious. Thus light industry has to be fostered. The expansion of light industry is also important in the sense that it mitigates the shortage of consumer goods that may occur in the beginning stage. The service sector has also to be focused, since its development has been delayed under the socialist system. As the migration and visits among the North Korean cities increase in the transition and reform stage, the demand for private services such as restaurants, hotels, and stores will also increase. Moreover, developing these service industries will decrease unemployment considerably because they have a great effect on job creation.

In the transition and reform stage, it is advised that North Korea join international financial institutions. By joining these institutions, North Korea will be able to receive various kinds of financial and technical assistance. In addition, this development will function as a signal that North Korea has become a responsible member of an international society, which will have positive influences on the environment for North Korea's transition and reform.

South Korea and International Community

Supports from South Korea and international society are requirements for North Korea's successful transition and reform. Although North Korea will firmly determine to promote economic transition and reform, it will be lacking sufficient internal resources and expertise. Above all, South Korea and international society must recognize that the peace, stability, and prosperity of Northeast Asia cannot be achieved without successful change in North Korea. Thus North Korea's neighbors have to be aware that their support is beneficial not only for North Korea but also for themselves.

The support can be categorized into two forms: making the environment favorable for North Korea's transition and reform and providing financial and technical assistance. The top priority for South Korea and international society is to make North Korea feel confident in maintaining its regime despite having abandoned military-first politics and activated transition and reform. The next leadership of North Korea, whether it is Kim Jung-un or not, is very likely to have a weaker power base than Kim Jong-il and thus must be anxious about whether he can secure his control over the state in spite of the expected negative effects of transition and reform. In the very beginning stage, therefore, South Korea and international society should conclude various diplomatic treaties with North Korea to guarantee North Korea's political system and indicate their strong intention to provide the maximum amounts of human and material assistance to help North Korea's transition and reform.

In the early stage of transition and reform, there may be serious economic difficulties and social disorder in North Korea due to the system change. So, economic assistance from South Korea and international society to stabilize the North Korean economy are vital. Additionally, the investments and assistance for North Korean firms and factories will be required to contribute to the construction of a self-sustainable growth base. In order to arrange and provide assistance to North Korea more efficiently, it will be desirable to form an "international council for assisting North Korea's transition and reform." This council can be under the existing Six-Party Talks and should welcome the participation of international institutions and other countries that are interested.

In this period, South Korea must prepare various measures such as tax exemption and financial incentives to encourage local governments, private firms, banks, and Nongovernmental Organizations (NGOs) to help North Korea. Forming sister relationships with the North Korean institutions should also be considered. The ministry of strategy and finance of South Korea must establish a blueprint with other related ministries to mediate systematically the financial and technical assistance to North Korea at the government level. South Korea can act as a major export market and investor for North Korea and will contribute greatly to smooth unification in the future.

It is also of great importance to secure human capital for a successful transition and reform. In this regard, it is strongly recommended that South Korea and international society establish local offices in North Korea to advise on various matters during the transition and reform stage. For instance, South Korea can organize an "inter-Korean consultation office for economic transition and reform" to provide advice and suggestions on issues such as privatization, price and trade liberalization, and stabilization. The Japen External Trade Organization (JETRO), the Asian Development Bank (ADB), and the International Monetary Fund (IMF) can also establish such offices. In addition, South Korea and international society should invite and train North Korean government officials, experts, and scholars who will be involved in North Korea's transition and reform and build training centers within North Korea to meet domestic demand.

Transformation Stage

North Korea

If the transition and reform is successfully completed, North Korea will enter the transformation stage. From the economic point of view, the transformation stage is the period when the market-economy system is settled down while market and plan coexist in the transition and reform stage. In the transformation stage, therefore, North Korea concludes the various tasks that have been promoted in the previous stage to achieve an advanced economy.

In this stage, North Korea starts privatization of large firms. As North Korea finishes privatization for all sectors, property rights belong completely to the private sector. The collective farming system should be disassembled and private farming should be legalized and encouraged.

Price liberalization should also be completed by lifting government controls on daily necessities and strategic goods to let the market replace all economic decisions. In addition, it is also necessary to abolish central planning legally. But careful inspection and mediation is necessary to prevent sudden and rapid inflation.

For the finance sector, a variety of openings should be pushed forward. The stock and bond markets must open and the foreign portfolio investment must be approved by liberalizing the capital market. And North Korea must allow the foreign banks, including the South Korean, to enter the market in order to earn more capital and acquire advanced financial techniques. Private insurance companies must be set free. And the floating exchange rate system must be introduced in this stage.

Trade should be liberalized fully. Free trade agreements with various countries will be a foothold of the economic development. It is especially desirable to conclude a "treaty for the economic community" with South

Korea to make inter-Korean economic relations much closer than those of any other countries.

South Korea and the International Community

In the transformation stage, what South Korea and international society have to do is to support North Korea to complete reform tasks successfully by adjusting financial and technical assistance to the changes and development of the North Korean economy. These steps imply that the assistance is cut in some areas but expanded in others. For example, the need to assist with basic daily necessities will decrease while assistance for privatization must increase as North Korea proceeds with mass privatization at this stage.

In the transformation stage, South Korea and international society must be concerned about the "reform fatigue" of the North Korean people that may accumulate as the system changes. The experience of the Eastern European countries shows that people tend to be negative toward economic reform due to high unemployment, decrease in real wages, and reductions in social welfare, although they fully support reform in the initial stage. Thus, South Korea and international society should be able to provide various kinds of assistance for North Korea to accomplish the economic transformation without backlash. Especially South Korea must support actively North Korea's economic transformation to create an opportunity for the coprosperity of both Koreas and unification on the Korean Peninsula. ■

■ References

North Korean Literature

Baek, Cheol-Nam. 2006. "Seon-gun-si-dae gyeongje-gu-jo-ui teug-jing" [The Characteristics of the Economic Structure in the Military-First Era]. *Economic Studies* 3.

Cho, Gil-Hyeon. 2007. "Gyeongje-sa-eob-eul jeon-mangseong-issge hae-na-ga-neun-geos-eun seon-gun-si-dae gyeongje-geon-seol-lo-seon-gwan-cheol-ui jung-yo-dambo" [To Promote Economic Projects Based on the Future Prospects Is the Important Security for Accomplishment of the Economic Policy Directions in the Military-First Era]. *Economic Studies* 4.

Cho, Ok-Sul. 2003. "Seon-gun-si-dae sa-hoe-ju-ui-gyeongje-bal-jeon-ui chu-donglyeog" [The Generating Power for Developing the Socialist Economy in the Military-First Era] *Economic Studies* 3.

Cho, Woong-Ju. 2005. "Seon-gun-si-dae gyeongje-geon-seol-lo-seon-eul cheol-jeo-hi gwan-cheol-ha-neun-geos-eun li sig sa-hoe-ju-ui-leul go-su-ha-gi wi-han hwag-go-han dambo" [To Accomplish the Economic Policy Directions in the Military-First Era Thoroughly Is the Firm Security for Sticking to Our Style Socialism]. *Economic Studies* 1.

Hwang, Cheol. 2006. "Gug-banggong-eob-wi-ju-ui gyeongje-to-dae-neun gangseongdae-gug-geon-seol-ui mid-eumjig-han dambo" [The Economic Foundation Based on the Military Industry Is the Credible Security for Building the Strong and Prosperous Country]. *Economic Studies* 1.

Kang, Cheol-Min. 2004. "Wi-dae-han su-lyeong gim-il-seongdongji-kke-seo ja-lib-jeog-min-jog-gyeongje-geon-seol-e-seo i-lug-ha-sin bul-myeol-ui eob-jeog" [The Immortal Achievements by the Great Leader Kim Il-sung in Making Self-Reliant National Economy]. *Economic Studies* 2.

Kang, Sung-Gil. 2002. *Seon-gun-si-dae-ui jo-gug-eul ga-da* [Going to the Homeland of the Military-first Era]. Pyongyang: Pyongyang Press.

Kim, Bong-Ho. 2005. *Seon-gun-eu-lo wi-lyeog-tteol-chi-neun ganggug* [Strong Nation through Military-first]. Pyongyang: Pyongyang Press.

Kim, Deok-Ho. 2004. "Jeon-lyeog, seog-tan, geumsog-gong-eob, cheol-do-un-su-leul apse-wo na-ga-neun-geos-eun seon-gun-si-dae-gyeongje-geon-seol-ui pil-su-jeog-yo-gu" [It Is an Inevitable Demand in the Economic Growth in the Military-First Era to Giving Priority to Electricity, Coal, Metal Industry, Rail and Transportation]. *Economic Studies* 2.

Kim, Hyeong-Seok. 2004. "Wi-dae-han lyeongdo-ja gimjeong-il-dongji-kke-seo balghyeo-ju-sin seon-gun-si-dae gyeongje-geon-seol-lo-seon-ui dog-changseong" [The Originality of the Economic Policy Directions in the Military-First Era Which the Great Leader Kim Jong-il Teaches]. *Economic Studies* 4.

Kim, Jae-Seo. 2004. "Seon-gun-won-chig-eul gu-hyeon-han sa-hoe-ju-ui-gyeongje-gwan-li" [The Management of the Socialist Economy Reflecting the Military-First Principles]. *Economic Studies* 1.

Lee, Dong-Gu. 2008. "In-min-gyeongje-ui gyun-hyeongjeog-bal-jeon-eun seon-gun-si-dae gyeongje-geon-seol-lo-seon-gwan-cheol-ui hwag-go-han dambo" [The Balanced Development of the National Economy Is the Firm Security to Accomplish the Policy Directions of the Socialist Economic Development in the Military-First Era]. *Economic Studies* 2.

Lee, Gi-Seong. 2003. "Wi-dae-han lyeongdo-ja gimjeong-il-dongji-kke-seo sae-lob-ge jeonglib-ha-sin seon-gun-si-dae sa-hoe-ju-ui-gyeongje-geon-seol-lo-seon" [The Policy Directions of the Socialist Economic Development in the Military-First Era Which Are Newly Established by the Great Leader Kim Jong-il]. *Economic Studies* 2.

Oh, Ju-Cheol. 2008. "Seon-gun-si-dae gyeongje-geon-seol-lo-seon-eun sa-hoe-ju-ui-gyeongje-geon-seol-ui bon-seongjeog-yo-gu-leul byeon-ham-eobs-i gyeon-ji-ha-go nop-eun-su-jun-e-seo sil-hyeon-hal-su issge ha-neun lo-seon" [The Economic Policy Directions in the Military-First Era Are to Maintain and Realize the Natural Demand for Constructing the Socialist Economy]. *Economic Studies* 2.

Park, Hong-Gyu. 2004. "Seon-gun-si-dae gyeongje-geon-seol-lo-seon-ui jeongdangseong" [The Justness of the Economic Policy Directions in the Military-First Era]. *Economic Studies* 1.

Park, Myeong-Hyeok. 2003. "Sa-hoe-ju-ui-gi-bon-gyeongje-beob-chig-gwa seon-gun-si-dae gyeongje-geon-seol-e-seo geu-ui gu-hyeon" [The Fundamental Rules of the Socialist Economy and Their Realization in the Economic Development in the Military-First Era]. *Economic Studies* 3.

Sohn, Yeong-Seok. 2005. "Wi-dae-han su-lyeong gim-il-seongdongji-kke-seo ja-lib-jeog-i-myeo hyeon-dae-jeog-in gug-banggong-eob-geon-seol-e ssah-eu-sin bul-myeol-ui eob-jeog-eun u-li in-min-ui ga-seumsog-e yeong-won-hal geos-i-da" [The Eternal Contributions to Build a Self-Reliant and Modern Military Industry by the Great Leader Kim Il-sung Will Be Permanent in Our People's Heart]. *Economic Studies* 2.

Yang, Ho-Nam. 2002. "Gyeongje-ganggug-geon-seol-e-seo gi-sul-gae-geon-ui jeol-bag-seong" [The Urgency of the Technical Progress in Building the Economically Strong Economy]. *Economic Studies* 1.

South Korean Literature

Han, Ki-Bum. 2009. *Bug-han jeongchaeg-gyeol-jeonggwa-jeong-ui jo-jig-haengtae-wa gwan-lyo-jeongchi: gyeongje-gae-hyeog hwag-dae mich hu-toe-leul jungsim-eu-lo* 2000–2009 [*The Organizational Behavior and Bureaucratic Politics in North Korea's Decision-making Process: Expansion and Retreat of Economic Reform 2000–2009*]. Ph.D. Thesis, Kyungnam University.

Kim, Bo-Geun. 2008. "Bug-han sang-in-gye-cheunggwa ja-bon-ui hyeongseong" [The Merchant Class of North Korea and the Formation of Capital]. paper presented at the seminar co-hosted by the Korean Association of North Korean Studies, Korea Institute for National Unification, and Institute for North Korean Studies at Korea University.

Kwon, Youngkyong. 2008. "7·1 jo-chi i-hu bug-han-jeonggwon-ui gyeongje-gae-hyeog · gae-bangjeon-lyag-gwa hyanghu jeon-mang" [The North Korean Strategy for the Economic Reform and Opening and Its Prospects After the July 1st Measures]. *North Korean Studies Review* 12, 1.

Lim, Soo-Ho. 2009. "Gimjeong-il jeonggwon 10 nyeon-ui dae-nae gyeongje-jeongchaeg pyeongga: 'seon-gun-gyeongje-no-seon'eul jungsim-eu-lo" [The Evaluation of Domestic Economic Policies during 10 Years of the Kim Jong-il Regime]. *Exim North Korea Economic Review,* Summer.

Park, Young-Ja. 2008. "2000 nyeon-dae bug-han no-dongja-ui no-dong-il-sang" [The Daily Lives of North Korean Workers in the 2000s]. paper presented at the seminar cohosted by the Korean Association of North Korean Studies, Korea Institute for National Unification, and Institute for North Korean Studies at Korea University.

CHAPTER 6

A Strategy for the Coevolution of North Korean Human Rights

Soo-Am Kim

Introduction

IN THE 1990s, FOOD SHORTAGES IN THE DEMOCRATIC PEOPLE'S REPUBLIC of Korea (DPRK) caused an impassable systemic crisis that forced Pyongyang to request emergency aid from the United Nations and the international community. During this period, a great many North Koreans crossed the border into China despite the DPRK regime's threat of punitive measures. From the defectors, stories about the deterioration of human rights in the DPRK began to spread to the international community.

North Korea is on a watch list under the UN's Human Rights Resolution, indicating the UN's concern for human rights conditions in the DPRK. The resolution "The Situation of Human Rights in the Democratic People's Republic of Korea" represents a decision of all member countries. Although it is not legally binding as resolutions of the UN Security Council are, adoption of this resolution nevertheless serves to put extra pressure on the North Korean regime, both politically and diplomatically. The North Korean regime views this resolution, publicizing human rights realities in the North, as a political conspiracy designed as a negative blow to North Korea's national security. In spite of the DPRK's rejection and denial, "The Situation of Human Rights in the Democratic People's Republic of Korea" works as a tool that has a significant impact on the North Korean regime's survival strategy.

Considering the current circumstances of international politics, *gangseong-daeguk* (a strong and prosperous state)—North Korea's survival strategy to maintain its supreme dictator—can be ridiculously erroneous in the twenty-first century. The new century's platform of international politics requires not just wealth and power, as that of the nineteenth century did, but also the soft values—such as knowledge, human rights, and preservation of the environment. The platform is thus becoming complex. For example, processes of development cooperation in the current international community no longer simply define the concept of "poverty" as "lack of wealth." Observers now pay more attention to comprehensive and multifaceted characteristics of "poverty," including "injustice" and "deprivation." These changes in the global platform make clear that another look at North Korea's survival strategy is necessary.

What does the North Korean regime need in order to come up with a survival strategy that is appropriate to this complex platform of the new age? A comprehensive reform in the DPRK's perception of human rights is a lynchpin for the North's future development strategy. If the DPRK just pretends to adopt the international standard of human rights and does not fundamentally transform its behavior, it will be difficult to expect future international support for North Korea.

In the future, it would be wise if North Korea adopted the human rights agenda insofar as it does not threaten the regime's safety and survival. North Korea must seriously consider whether it could survive in the twenty-first century should it persist in approaching human rights issues as a threat to the regime. In other words, Pyongyang has no other option but to agree to human rights concerns and search for survival strategies in tandem with the international community. This paper looks at the implications of the human rights agenda for North Korea's survival strategy, and in addition, it analyzes and suggests a direction in which North Korea must approach human rights issues in order to succeed in its prosperity strategy.

Human Rights Awareness in the Military-First Era: Content and Evaluation

Perceiving the Human Rights Issue as a Threat to the DPRK Regime's Security

In the 1990s and in the military-first era, the DPRK's perception of human rights was established based on the fear of threats from the outside.

The fall of communism and Pyongyang's own evaluation on the causes of the fall greatly affected the way Pyongyang understands human rights. In the 1990s, when the Soviet Union and communist states in Eastern Europe collapsed, North Korea was faced with a grave threat to its national security. The North Korean regime believed that the expansion of freedom and human rights and the subsequent escapes of the residents en masse were the main causes of the fall of communism. The DPRK regime took a special caution to restrict the inflow of information, which it believed to have shifted public perception. For this reason, the international community's criticisms of human rights violations were interpreted as a security threat. To support this assumption, the regime spoke of the war in Iraq as an exemplary case that revealed the covert intention of the United States and the international community to topple the regime in North Korea.

North Korea misunderstood the international community's strategy as one of bringing down the socialist system and regimes and justifying its actions in the name of human rights. The North defines the world order as a battle between imperialism and self-reliance, and human rights as a weapon of imperialism. The DPRK argues that the ambition of the imperialists, which is to dominate the world, never changes but only their methods do. In short, human rights are being used as a tool of the imperialists in dominating the world, in the eyes of the North Korean authorities (*Rodong Sinmu*n August 17, 2007).

The North Korean regime particularly emphasizes that the "human rights attack" is a strategy to ruin socialism and believes that this strategy must be countered considering the context of the system's security. To put it another way, the Western way of public discussion of human rights is nothing but a nominal reason that Westerners employ in order to bring down socialism. At the core of that strategy, North Korea points out, lies the proliferation of human rights awareness and support for antiregime forces—the final goal of which is the fall of socialism and its core values such as solidarity and collectivism. According to Pyongyang, the Soviet Union and other socialist states in Eastern Europe fell because they failed to see through the toxic intentions hidden behind the "human rights attack," and therefore no effective management measures were developed against such a strategy. Looking back on this historical lesson, North Korea claims that it must prepare a fierce ideological front line against Western tactics for the protection of the socialist system. In short, the North Korean regime is taking a security approach in dealing with human rights issues (Kim 2007, 72–73).

Pyongyang argues that there is not one absolute standard for human rights. Due to the differences in cultures, there cannot be a single standard for human rights that is applied to every state. Therefore, the insistence of Western states to apply "their way" of human rights to North Korea, regardless of the differences, can only be interpreted as a strategy to conquer the world. North Korea also applies this argument against globalization. For North Korea, globalization is another strategy to conquer the world based on Western values, in particular, American values. North Korea calls it "homogenization." Through this homogenization, North Korea asserts, Western states are trying to turn the entire globe into one free world and then subordinate and assimilate all the people to their system. In particular, North Korea criticizes the United States for trying to "Americanize" other states by setting its standard of human rights as a global standard. In this way, the North Korean regime perceives Western and American diplomatic strategy to impart the values of human rights as a threat to their system (Kim 2000, 329–333).

Pyongyang sees that it is the major target of the "human rights attack" and looks at the expansion of global interests in its human rights issues from a security point of view. The United Nations has viewed the human rights situation in North Korea as a pressing and serious matter and has requested improvement through a special process. For two consecutive years in 1997 and 1998, the UN Sub-Commission on Human Rights passed a North Korean Human Rights Resolution. When the North Korean regime did not show a definitive will to make progress, the UN Commission on Human Rights (UNCHR) adopted the North Korean Human Rights Resolution for three consecutive years from 2003 to 2005. Especially in 2004, the UNCHR appointed a Special Rapporteur on the situation of human rights in the DPRK, and launched an investigation.

The military-first regime's idea on human rights is well implicated in its attitude toward the North Korean Human Rights Resolution, which is continuously adopted by the UN Human Rights Council and General Assembly. After the UNCHR passed the North Korean Human Rights Resolution in 2003, it virtually replaced the UN Human Rights Council and undertook the role of overseeing North Korean human rights. Even under the new system, the resolution was passed for four consecutive years from 2008 to 2011. In 2005, it was agreed that the matter would be passed on to the UN General Assembly should North Korea continuously refuse to accept the mandate of the resolution passed by the UNCHR. Therefore, from 2005 to 2010, the North Korean Human Rights Resolution was passed in the UN General Assembly. Unlike the Human Rights Resolution passed by the UN Security General, the resolution of the UNCHR does not carry coercive power. Nevertheless, these resolutions aimed at several states that violate human rights put pressure on North Korea, politically and diplomatically. The North Korean authorities persistently refused to accept the resolution and criticized it as a tactic to strangle North Korea. In addition, the authorities vehemently blamed the United States for playing a predominant role in writing the resolutions and said that other states such as the European Union and Japan have jumped on the bandwagon in order to overthrow the DPRK system.

As reflected in North Korea's stance toward the UN resolutions, the North Korean authorities regard the human rights policy of the United States as the primary threat to its security. The North Korean regime, hence, set up its security strategies with the foremost concern for the human rights assault. What made the weight of the matter even heavier was the Bush administration's public labeling of North Korea as an "axis of evil" and an "outpost of tyranny." To make matters worse, U.S. high officials, including George W. Bush, directly blamed North Korean leaders. North Korea regards such remarks as a challenge to its totalitarian dictatorship. As the United States pushed for a human rights policy against the North through its own legislation, North Korea's perception of human rights as a threat has deepened. In the North Korea Freedom Act of 2003, several provisions regarding weapons of mass destruction were included, and it fueled further sentiments of threat in Pyongyang. Reflecting on such criticisms, an amended bill was passed in 2004, yet North Korea perceived that even the adoption of the document itself posed a threat to the regime's security. Unlike the Iraq Liberation Act, the North Korea Human Rights Act of 2004 does not stipulate a regime change. However, North Korea's original perception of the human rights policy persisted.

Perceiving Human Rights through the Lens of National Sovereignty

North Korea's perception of human rights as a security threat brought about a counterstrategy that emphasizes national sovereignty in the military-first era. It has three specific distinguishing features.

First, the unique path of communitarianism, which is differentiated from that of general socialism, is the key factor that invokes the North Korean regime's approach to human rights issues from the security point of view. The North Korean authorities believe that "our way of Socialism" is the system that genuinely realizes social collectivism. Also, North Korea's socialism differs from past socialism. The North Korean Constitution states, "in the Democratic People's Republic of Korea the rights and duties of citizens are based on the collectivist principle, 'One for all and all for one'" (Article 63). The North Korean way of socialism has special distinguishing characteristics: the revolutionary *Suryong*, or Great Leader, ideology, absolute loyalty to the dictator, the ideology of "one organism" that the state is one body and the *Suryong* is the head, and the theory of "one large family of socialism" (Kim 1998, 365–366). Therefore, the idea of human rights, which designates individual citizens' *Juche*, or self-reliance, inevitably interferes with North

Korea's unique communitarianism, and the North Korean leaders perceive this issue as a serious security threat to North Korea.

Second, the principle of extreme hierarchy and exclusivity prevents North Korean leaders from accepting international human rights demands. Currently in North Korea, rules that justify the exclusion of certain classes are being actively used for the security of the regime. In particular, there is a tacit agreement among the leaders that reckless repression of anti-*Suryong* forces is rightly justified: "We do not hide our stance on political matters, nor on the human rights issue. Socialistic human rights do not extend to the undeserving elements, including hostile and impure ones who infringe on the interests of the People." (*Rodong Sinmun* June 24, 1995)

Thus the security of North Korea is based so strictly on the principles of hierarchy and exclusivity that it is difficult to expect North Korea to accept a general application of human rights.

Third, although North Korean society prioritizes politics as general socialism does, it has a unique governing mechanism that differentiates itself from the others: absolute loyalty to the *Suryong*. This, in fact, is fundamentally hampering North Korea from establishing governance. In the North, super-legal measures such as the 10 Principles of "unitary ideology," tenets, words, and proclamation take precedence over other rules in regulating people's behavior. In such ways, *Suryong*-oriented principles have negative impacts on establishing governance and make North Korea respond in an inflexible way to the demands of the international community. Worst of all, and most inhumane, is the system known as *Yeon-jwa-je* (or guilt by association). This system allows the North Korean authorities to punish not only the person who commits illegal activity but also his or her family. Furthermore, people's social background is inherited from one generation to the next under this system. The reason why North Korean residents fear *Yeon-jwa-je* is that not only the criminal gets punished but the entire family can be harmed because of that person. The residents testify that they do not dare to express their complaints toward the regime, due to the fear that their families might fall victim to this system (Lee et al. 2009, 94).

Looking at the human rights issue from the regime's security perspective, strong connections between the human rights of the military-first era and military-first strategies can be seen.

Since the 1990s, the North Korean authorities have been developing the idea of military-first in order to counter the human rights issue. That is, the North Korean authorities are applying cultural relativism to proclaim "our way" in the realm of human rights. The authorities argue that there is no standard application of human rights. Therefore, for the international community to ask for the "Western way" of human rights cannot be justified. According to the North Korean regime, the "Western" standard of human rights does not fit their state, and instead, what the North Korean people prefer—"our way of human rights"—is the best fit for North Korea (Kim 1996, 371). But their argument does not lay out any specifics; it does not contain any details other than the claim that the North Korean people are enjoying human rights under "our way of socialism."

One of the reasons the North Korean regime regards the human rights issue as a threat to their regime is that its human rights concept is directly connected to its founding principle. The uniqueness of the North Korean system comes from the fact that it applies both *Juche* and the military-first ideology to its concept of human rights. This is well reflected in the North Korean constitution, which was modified at the first session of the Twelfth Supreme People's Assembly (SPA) held in April 2009: "The DPRK considers the *juche* idea and the military-first, which are people-oriented worldviews and revolutionary ideas for achieving the independence of the popular masses, as the guiding principles of its activities" (Article 3).

What makes North Korea stay rigid and inflexible toward international demands regarding human rights, and also worsens the problem, is the *Juche* ideology. In North Korean literature, human rights are defined as "rights of life for a social existence with independence." Also, the North's authority claims that the way to realize "our way" of human rights overlaps with realization of the *Juche* ideology. In short, human rights in North Korea is closely interrelated with *Juche* ideology. North Korea's human rights are unavoidably linked to the social ideology of "socio-political organism," meaning that only when the *Suryong*, the Party, and the people are united as one living entity can there be assured human rights in North Korea. In other words, when members of the society work for a common goal and shared fate under the right leadership of *Suryong*, the society can truly realize "independent human rights." Under this logic, however, human rights become a blessing that the *Suryong* bestows on the people, not an inherent and natural right of human beings. Therefore, by the revolutionary *Suryong* ideology, the concept of human rights degenerates into a mere sub–concept of "unitary ideology" or a justification of the system (Seo 2007, 142–153). There is a new tide of human rights in the changing world of the twenty-first century. Yet North Korea remains rigidly bound to its worn-out dogma.

Chained by its *Juche* ideology, the North Korean concept of human rights is also closely tied to the military-first strategy. As with other socialist states, North Korea also responds to international pressure with the idea of national sovereignty and nonintervention arguments. However, North Korea innately differs from the other socialist states in that it looks at human rights as a matter of "state rights," not just national sovereignty. To North Korea, imperialists have a plan to interfere in other states' domestic affairs and break down the system, or topple the regime in the name of human rights. In particular, North Korea claims that the United States is trying to project this imperialistic strategy in their land. The war between the United States and Iraq has deepened this distrust and doubt the North has had toward the United States. North Korea criticizes the United States for justifying its invasion with bogus reasons such as the "freedom" of the people from a "non-democratic and repressive regime" and says that the people of North Korea should never forget the fact that the loss of state rights means no human rights and no existence at all (*Rodong Sinmun* August 19, 2006).

On the basis of this "evidence," the North Korean authorities continue their argument that human rights are essentially the same as state rights. Namely, since the Western demand for human rights reform is nothing but an attempt to topple the regime, protecting state rights can be equal to the protection of human rights. Hence, according to the authorities' argument, the protection of state rights must proceed, in order to guard human rights in a genuine sense. North Korea's way of looking at human rights—regarding human rights as a threat to the system's independence—has borne this dangerous conclusion (*Rodong Sinmun* August 17, 2007).

When this argument is made, the concept of human rights is combined with the military-first strategy. For the North Korean authorities, who firmly believe in imperialism's belligerence, it becomes inevitable to build military forces that can guarantee human rights, and therefore state rights, in the midst of increasing pressure from the international community. The North Korean authorities argue that strong state power, namely military power, is not only a necessary condition for human rights but is also the best political method to protect genuine human rights. Therefore, the North puts forth the philosophy that the military-first approach is the most trustworthy way to achieve human rights, and that military-first politics are, in fact, the politics for human rights (*Rodong Sinmun* September 13, 2007). Ultimately the two elements, the "revolutionary spirit of soldiers" and the "spirit of escorting the Great Leader," ossify the response strategies of the North Korean authorities toward the human rights issue. Through military-first politics, the North Korean authorities have set up the Korean People's Army (KPA) as a model for the whole society in order to prepare it as a political force with a highly revolutionary spirit. This revolutionary army regards revolution, organization, unity, and rules as central tenets (Jeon 2004, 75–81), and it is trying to impart those values to the people and the society. Especially, through expanding the spirit of soldiers, including the "spirit of guarding the Great Leader with determination," "the spirit of perfectly completing goals," and the "spirit of heroic sacrifice," the North Korean authorities are trying to build human robots armed with absolute loyalty and unconditional obedience. There is no room to establish personal identity under such conditions.

The military-first strategy works as a key element in determining the North Korean authorities' policies regarding the human rights issue. Also, the military-first strategy is a pivotal element in determining North Korea's development strategy, and the development strategy in turn is a lynchpin factor that influences the quality of life and human rights of North Korean residents. Military-first politics, according to the North Korean authorities, form a political system that prioritizes the military in the state's agenda. More accurately, it is a politics that prioritizes the entire area of the military such as troops, defense industry, war, and all other areas related to national defense as the most important industry of the state. Hence, military-first politics seeks to open the gate of *gangseongdaeguk* through the means of defense industry and the military. Should North Korea persist in the development strategy that is grounded in military-first politics, the uneven distribution of resources will unavoidably undermine the natural rights of the people of North Korea.

The North Korean authorities approach the human rights issue from the standpoint of security of the regime and the system, a perspective deeply rooted in *Juche*, military-first, and cultural relativism. In an attempt to respond to the international pressures with the UN at the center, the North Korean authorities have undertaken some limited amendments to the constitution but only as a covert strategy to calm the pressure. In 2009, the authorities added an article that guarantees the human rights of the North Korean people. Article 8 of the revised constitution states, "The state shall safeguard the interests of, and respect and protect the human rights of the working people, including workers, farmers, soldiers, and working intellectuals, who have been freed from exploitation and oppression and have become the masters of the state and society." In 2003 the government passed a law to protect people with disabilities and since 2004, criminal law and the Criminal Procedure Code were revised as a part of efforts to improve human rights conditions in North Korea. In addition, the authorities have been allowing limited cooperation, for example, by turning in national reports and dispatching representatives to be investigated according to the four major international human rights instruments North Korea has joined. As explained earlier, there have been partial changes in North Korea's attitude toward human rights at both the legal and the diplomatic levels.

Gradual Decline and Search for an Escape Route

The North Korean authorities are in a search for a survival strategy in their own way, which is to open a gate of *gangseongdaeguk* by 2012. In the short term, the North will seek to establish a survival strategy under the *Suryong* system. But the human rights problem will play a significant role as North Korea works for that survival strategy; the North will not bother to amend its view of the human rights issue as something intertwined with the security of the regime and its system. Only a few limited revisions will be made merely to dilute international pressure. Yet it will not be easy to garner international support without fundamentally shifting the North's view.

Although the North Korean authorities will try to meet international pressure on human rights by amending a few laws, the international community will want to see how those legal changes actually help the North Korean people to gain their rights. North Korea will put forth its amended constitution and the criminal procedure code as decisive evidence that the authorities guarantee the human rights of the people. However, without changing the fundamental tenets of this politics-dominated society, the superficial legal revisions will merely produce secondary effects. As long as the core function of the criminal law and the Criminal Procedure Code is to protect national sovereignty and the socialist system, the human rights article will be a secondary tool with inherent limitation from its inception. Also, in North Korea, a politics-dominated society, the principle of hierarchy and the political line of the mass will be considered fundamental elements rather than human rights principles.

A number of elements in the North Korean legal provisions are still subjected to international criticism, and one of the most inhumane systems, the public trial, receives focused attention from the international community. Open trial in North Korea is not used to guarantee the due process of law and human rights. Instead, it is used to evoke fear among the people and thereby maintain the social order. Although North Korean law guarantees the right to hire an attorney, the government nominates the attorney for the defendants. Another inhumane element criticized by the international community is the public execution system. Public trial and execution are both inhumane because of political manipulation of individuals in order to control society. Also, although the North Korean authorities will try to advertise the improvements in their human rights system, its application and effectiveness are bound to the limits of the monolithic ruling system. Should North Korea persist in its political prisoner camp and public execution system as well as its approach to human rights issues for the protection of the *Suryong* monolithic ruling system, progress in human rights awareness can hardly be expected. The current situation, in which even the domestic laws are not properly enforced, is likely to continue due to the inhumanity in the constitution and low human rights awareness among the North Korean legal workers. Until legal revisions in North Korea produce practical outcomes, conflicts with the international community over human rights will be heightened. Also, as long as North Korea's nuclear issue is deadlocked, the human rights problem will act as a barrier to any possible cooperation between the North and the international community. The North Korean authorities have to be aware that human rights matters will hamper it from building *gangseongdaeguk*, should it continue to pursue its inhumane practices.

North Korea's dogmatic refusal strategy to the international demand for human rights can bring some short-term security to the *Suryong* monolithic ideology, but in the long run, it will be detrimental to the regime's security as well as to its survival strategy. The more refusal strategies North Korea employs on the human rights issue, the more attention it will draw from the international community, hence put more pressure on the North Korean authorities. The international community's interest in the North Korean human rights problem is well reflected in the Universal Periodic Review (UPR) that the UN Human Rights Council newly adopted in order to enhance the surveillance capability of the human rights issue. During the general Universal Periodic Review held on December 7 in 2009, more than 60 member nations applied to speak on the floor, though only 52 nations were able to speak due to time constraints. This number, compared with participation rates in other nations' Universal Periodic Reviews, is a clear indication of the great interest that UN member nations have in North Korea's human rights problem. Even ranks of the speakers differentiated the members' interest in North Korea from the rest in the general Universal Periodic Review of other nations; there are even cases when remarks were given by interns, but for North Korea, ambassadors themselves spoke. Notably, Special Envoy for North

Korean Human Rights Issues Dr. Robert R. King spoke for the United States. Considering such international interest in North Korean human rights conditions, the issue is likely to be set as an official agenda item for the UN in various ways within the UN Human Rights regime, unless the North immediately shows a sign of shifting its attitude on the issue.

Of 167 recommendations suggested in the general Universal Periodic Review of December 2009, the North Korean authorities rejected 50 recommendations and took note of 117. The UN Human Rights Council adopted an official report based on the recommendations outlined in the "UPR Working Group Report on North Korea." The next UPR on North Korea will be held in 2013. Until then, the international community will intensify its pressure for better human rights conditions in North Korea and continue to watch over the authorities there. During this time, North Korea's reaction toward the recommendations, especially the noted ones, will decide the future strategies of the international community to a large extent. North Korea's reaction toward the general UPR, and the international community's reaction toward the North's reaction, will be a key factor for the success of the survival strategy that North Korea is pursuing.

If North Korea persistently pursues a refusal strategy against the UN Human Rights Council, the UN General Assembly, along with the follow-up measures of the general UPR, will tighten its monitoring on behalf of the international community. The North Korean authorities must keep in mind that the international community evaluates North Korea's willingness to improve human rights conditions based on its attitude toward special procedures and technical cooperation in human rights. The North Korean authorities do not recognize the existence of the UN Special Rapporteur for North Korean Human Rights and refused to allow its visit, because the Rapporteur was nominated by a UN Human Rights Resolution. Similarly, just because it is included in the North Korean Human Rights Resolution, the authorities have been refusing to cooperate with the United Nations Office of the High Commissioner for Human Rights (UNOHCHR). The North's uncooperative behavior toward the activities of UN human rights organizations will only foster the constant adoption of the North Korean Human Rights Resolution in the

UN Human Rights Council and the UN General Assembly. Also, repeated turndowns on UN Special Rapporteurs and international human rights NGOs'requests for visits, without the North's having a self-monitoring system, will only aggravate the conflict between the North and the international community. If the adoption of the North Korean Human Rights Resolution is extended over the long term, it will negatively influence North Korea's ability to gain international support, which is, again, pivotal to North Korea in building *gangseongdaeguk.*

If North Korea keeps reacting against the human rights issue from the perspective of security and threats, more and more advocate activities to regarding negative public opinion will sprout from human rights NGOs around the world. Especially, demands to punish and hold the North Korean leader to account for human rights abuses will further increase, with NGOs at the center. Furthermore, despite the low chance of attainability, international voices asking the International Criminal Court (ICC) to sue Kim Jong-il on charges of inhumane crimes against humanity will gain momentum. Already, the U.S. Committee for Human Rights in North Korea and other human rights NGOs have defined in the Human Rights Report that violations of human rights in North Korea are inhumane crimes and suggested some ways to sue the violators in the UN Security Council and the ICC. The Crimes Against Humanity Investigation Committee, a civil organization coalition, was established in order to sue Kim Jong-il in the ICC. After 2010, there will be enhanced activities to induce more active intervention of the ICC in the investigation of North Korea's violations of human rights, using a preliminary examination system underscored in article 15 of the ICC statutes. Such intervention could yield results quite contrary to North Korea's goal, protection of the *Suryong*, should North Korea insist on its long overdue approach to the human rights issue. Apart from attainability, if the effort to sue Kim Jong-il and the solidarity of the international community intensify, they will pose a significant threat to the security of the North Korean regime. Also, international NGO-led suggestions to make the leader bear responsibility will negatively impact the North's survival strategy, which requires cooperation of the international community.

With the nuclear problem deadlocked, the human rights agenda will not only be a barrier to the normalization of U.S.-DPRK relation but will increase the deterioration of relations even further. In 2004, when the United States enacted the North Korea Human Rights Act, the North Korean authorities criticized the act as a political ploy to topple the North Korean system with the nuclear issue on the one hand and the human rights issue on the other. If the North Korean authorities do not show signs of changing and persist in their ways, the United States will have to resort to tougher diplomacy on human rights, based on the North Korea Human Rights Act. The United States Congress extended the act, which was effective until 2008, for four more years through the North Korea Human Rights Reauthorization Act. Through this act, the part-time position of special envoy for North Korean human rights issues became full-time, and this in turn has laid a foundation for professional planning and implementation in the future. During this stage, if North Korea remains uncooperative toward North Korean human rights envoys, the human rights issue could damage U.S.-DRPK relations. There is also a possibility for the United States to adopt the North Korean human rights issue as an official agenda item for either bilateral dialogue or in the Six-Party Talks. In addition, the United States would probably extend the North Korean Human Rights Act again if conflicts over human rights and nuclear weapons continue with no viable solutions. The North Korean authorities must be reminded that their dogmatic view that the universal value, which in this case is human rights, is an extreme threat to their security will not be beneficial to their relations with the United States nor to their attempt to gain international aid. Hence, the authorities are likely to be badly mistaken by persisting in their current approach, if they want to build *gangseongdaeguk* by 2012, as they have proclaimed.

If the North Korean authorities do not act for the improvement of human rights conditions, the United Nations, individual states, and nongovernmental human rights organizations will form an intertwined network at all levels regarding the human rights problem in North Korea. Especially within South Korean society, more and more activities to improve human rights conditions in North Korea will emerge with the support of international solidarity.

In South Korea, a currently proposed bill to improve human rights conditions in North Korea will be passed if the North does not show any signs of improvement. According to the proposed bill, the Ministry of Unification has designed a basic plan to achieve better human rights conditions in North Korea every three years. Also, the special envoy for North Korean human rights will be appointed, and a North Korea Human Rights Foundation will be established for more systematic investigation of the deplorable human rights conditions in North Korea. The foundation will provide financial backup for civilian groups that work to improve conditions, which will in turn expand the public's voice in South Korea on the issue as advocacy groups and monitoring activities become galvanized as a natural consequence of financial support. More and more information inflow through the media will follow. Coupled with the human rights bill, it will add new momentum to advocacy activities in South Korea. Furthermore, when the Special Envoy for North Korean Human Rights is appointed through the human rights bill, it would be reasonable to expect enhanced cooperation between South Korea, the United States, and Japan. If North Korea maintains its passive and negative attitude toward issues related to POWs, DPRK abductees, and separated families, public opinion in South Korea is likely to become louder and ever more negative. A vicious circle of invigorated human rights activists in South Korea, North Korea's complaints against them, and more aggravation of South Korean public opinion will emerge and hamper North Korea from pursuing a path toward *gangseongdaeguk* in the long run.

The international community has been accumulating various ways to maximize the effect of development cooperation to North Korea. However, in order to receive such aid, a receiving country must make some change in its policies. Therefore, the North Korean authorities must change their policies and system in order to receive the massive amount of development cooperation they need in order to build *gangseongdaeguk*. In the process of change, however, North Korea should abide by international principles such as the five principles of the Paris Declaration on Aid Effectiveness and Poverty Reduction Strategy Papers, which include significant human rights elements. Therefore, if North Korea maintains its isolated system centered on *Suryong*, even though it adopts human rights and open policies to a limited degree, a large sum of international development cooperation is unlikely, a natural consequence of which is a smaller and smaller likelihood that North Korea will achieve its goal of *gangseongdaeguk* by 2012.

As long as the North Korean authorities keep pursuing the current form of the *Suryong* system, using imbalanced development strategies, resource distribution that discriminates, and control mechanisms, conflicts with the North Korean people will deteriorate and become a barrier to the security of the regime. Unless the authorities put an end to military-first politics and shift to a fundamentally open policy, the people of North Korea will inevitably be faced with survival problems on their own. Amid a weakening economy and resource imbalance, the North Korean people are seeking ways to survive through the market. Unfortunately, however, the North Korean authorities have set the military-first and *Juche* ideologies as guiding principles and employed coercive measures to control the lives of the people, and all of them have proved to be counterproductive to the regime. It is likely that the conflict between the regimes, persisting on coercive and ineffective measures, and the people, who want survival, will grow larger, and the number of defectors fleeing from the regime for survival will continue to soar. Regarding the swelling number of defectors, more disputes between the North, South Korea, and China are anticipated. In addition, if the problem of defectors becomes a national issue, the North Korean authorities' effort to punish defectors will draw international criticism.

Many North Korean defectors testify that despite the authorities' attempt to control the people, after the economic downturn in the 1990s, increasing amounts of outside information are flowing into the communities of survival-seeking people, the number of whom is growing at a fast rate. An increasing number of people are watching South Korean videos and television and radio programs in secret, as well as Chinese ones. With the border area as the center of circulation, South Korean dramas and movies are spreading among the residents. There is a limitation to the North Korean authorities'coercive measures because the inflow of outside information comes by unofficial methods, and more and more people have begun to compare North Korea with the other countries they see on TV and radio. Diffusion of such information will make the North Korean regime uneasy about its security. Therefore, instead of negative measures such as filtering and coercion, the North needs to take positive approaches such as improving the life quality of residents through an open-door policy. The North Korean authorities must realize that if they do not take immediate measures to improve human rights conditions for the regime's security, the safety of the regime cannot be assured.

Coevolution Strategy for Advanced North Korean Human Rights

Stage of Implementation and Reform

North Korea

With abandoning the nuclear program as a start, through escaping from its political isolation and financial support, North Korea must fundamentally redesign its strategy of survival and prosperity in conformity with the international order of the twenty-first century. In order to move beyond mere survival and attain prosperity in the international political platform that has complexity as its character, the North Korean authorities have to recognize the importance of soft power, such as knowledge, environment and human rights. About human rights in particular, North Korea should break out of the state-centered ideology that has grown out of the principle of sovereignty and recognize the complexity of the issue, in which various actors—the United Nations, individual states, and international NGOs—are complexly involved and interrelated. Regarding the human rights issue, an important universal value, North Korea should flexibly interact with other involved entities—UN human rights organizations, UN member nations, and nongovernmental organizations.

The North Korean authorities must realize that improvements in human rights do not pose a major threat to the security of the regime. They must strive to form a new strategy so that the two elements do not interfere with each other in a negative way. Two transitional steps should be taken for this new strategy. First, the major concern of the North's leadership is that the regime might collapse if the values of human rights slip into society. But nowhere in the world has awareness of human rights matured in a short span of time. Only through reciprocal communication and interaction with the international community for medium to long periods of time can human rights awareness truly mature in a society. Hence, the North's current strategy of dealing with human rights from a security perspective requires a major overhaul, from what the North defines as reaction to "aggression" to that of fostering communication and cooperation with the international community. The North should take a prudent yet steady approach—such as conversation and cooperation with the international community and domestic measures for a step-by-step implementation—to the human rights agenda.

More specifically, the North's strategy of rejecting the international demand for human rights by using the principle of extreme sovereignty (especially, national sovereignty) must be altered. If North Korea persists in its long-held political dogmatism that human rights can only be assured through the military-first strategy, it will not be able to step forward toward a successful survival strategy with the support of the international community. Such dogmatism must yield its place to other strategies. Only then will North Korean authorities be able to take into consideration the five-phase dynamic model of human rights change that Thomas Risse and Kathryn Sikkink have developed. According to Risse and Sikkink, the first stage of change in a violating state is a *repressive stage*. The next phase is a *state of denial*, in which the accused government denies the demands of the international community for improved human rights. The first reaction of most governments that are accused by the international community is denial, a fundamental challenge to the legitimacy of international human rights norms. The third stage is one of *tactical concessions*. As the pressure from the international community intensifies, the repressive state attempts partial concessions such as releasing political prisoners and allowing more political activities. In most cases, such concessions tend to remain tools to pacify international criticism without genuine change in awareness. After that comes the *phase of prescriptive status*, in which the legitimacy of human rights standards is no longer a matter of controversy. The final stage of the model emerges when the behavior becomes *rule-consistent behavior*. In this stage, human rights norms become a common practice and the rule of law is upheld (Risse and Sikkink 1999, 1–38; Lee 2008; Heo 2008).

If this model is applied to North Korea, the North's authority can be placed in a stage between repression and denial, while partial concessions are strategically employed. In other words, as generally perceived, the North Korean authority would not rigidly pursue a strategy of denial but make strategic concessions to international pressure. The North's strategic concessions can be evaluated on two levels. From the security point of view, the authorities' strategic concessions have not reached political areas such as political imprisonment and public execution. In other words, strategic concessions are not extended to the political field. Only in areas that do not influence politics, such as domestic law and international cooperation, are limited concessions being made. However, such limited strategic concessions have limits in mitigating international concern about the North's violation of human rights. The North's concessions on human rights issues must be extended to the area of politics if North Korean authorities want to root out the negative elements that this agenda places on their survival strategy.

In order to reduce international concern over the North's mistaken approach to human rights, the authorities in the North must expand their strategic concessions, not increase denial. Strategy must be changed at two levels.

First, in the beginning stage of the coevolution, North Korea must answer to the demands of the international community so that there is no need for adoption of the North Korea Human Rights Resolution by either the UN Human Rights Council or the General Assembly. For this, in an attempt to modify its human rights policies, Pyongyang ought to strengthen cooperation with UN human rights organizations and openly accept the advice of the Human Rights Resolution, the UPR, and Concluding Observations of the UN committees based on international human rights instruments. In particular, North Korea should selectively allow visits of the Special Rapporteur, which is one of the UN's Special Procedures. On the basis of such cooperation, the authorities have to evaluate the visits of the United Nations Special Rapporteur with a forward-looking attitude. Also, the North needs to expand its cooperation with the UN committees based on international human rights instruments and allow members of the committees and international human rights NGOs to visit North Korea. Such expansion of open cooperation will render the adoption of the North Korea Human Rights Resolution unnecessary. In addition, North Korea should redirect its strategy to expand cooperation in the area of human rights to include more conversation and technical cooperation. It must join the UN Convention against Torture and Other Cruel or Inhuman or Degrading Treatment or Punishment as the international community is asking and continuously work toward expansion of cooperation with the International Labour Organization (ILO).

Second, North Korea must attune its strategy with that of the international community so that it can be imparted to individual states. Especially during the process of resolving the nuclear problem, the North Korean authority needs to lay the groundwork for a talk with the United States in regard to the human rights issue. The North Korean regime should persuade the decision makers and the public in the United States that the North Korean Human Rights Act is no longer needed. That would in turn put pressure on the U.S. government to abolish the act. During this process, the North Korean authorities ought to improve cooperation with other states interested in its human rights, including those in the European Union, Australia, and Canada. Also, it has to seriously consider cooperation with South Korea in regard to human rights. More cooperation with individual states could help North Korea to find a better position in its attempt to gain larger amounts of financial aid from the international community. In order to successfully pursue a strategy befitting the international reality of the twenty-first century, the North Korean authorities should make clear to the international community that they want to pursue at least four stages of the five-phase dynamic model that Risse and Sikkink have developed. In other words, the North must do more than mere strategic acceptance and change its strategy gradually to the *phase of prescriptive status*. If such modification is made, positive international conditions for the twenty-first century survive and ultimately prosperity will follow through reformation, an open-door policy, and developmental cooperation.

If the authorities want to move beyond coercion, rejection, and partial and strategic concessions and seek security for their regime and a twenty-first century survival strategy through cooperation with international society, they must escape from their extreme cultural relativism. In particular, the North Korean authorities should gradually modify their cultural relativism regarding human rights. The beginning of the modification can be categorized into four classes, according to Jack Donnelly. First is radical relativism, namely, the view that only culture, history, and economy are the genuine origin of all values, and there is no such thing as inherent rights that every human being has had bestowed equally from birth. The second category is radical universalism, which regards all values as essentially universal, unaltered by cultural or historical factors. Next is strong relativism. This category sees that in principle, human rights are determined by culture and other environmental factors, though not entirely. The "universality" of human rights can have meaning as a device to check certain values, but this idea is based on the change of values and relativism. Finally, in weak relativism, universal human rights are viewed as a standard, but historical and cultural diversity can be factored in an attempt to realize rights. However, this idea ensures that even various methods of implementation must be restricted so that the core contents of rights do not get distorted (Donnelly 2002, 73–76). Of the four categories, the North Korean authorities' beliefs can be located somewhere between radical relativism and strong relativism. In the meantime, however, the authorities must abandon their nuclear program, transition to an enlightened *Suryong* governing method, and seek survival through reform and open-door strategies. During this transition period, the authorities have to gradually shift their stance on human rights from radical relativism to strong relativism, and eventually, to weak relativism.

In order to change their stance from strong relativism to weak relativism, the North Korean authorities must first get rid of the "our way of human rights" principle. As long as North Korea continues to approach the human rights issue from a black or white, "our way" or "Western way" perspective, it will be extremely difficult to form cooperative ties with Western advanced states, including the United States and the members of the European Union. Only when the North Korean authorities drop this worn-out perspective can they participate in talks with the West and work cooperatively with them. Therefore, the North Korean authorities must reconstruct their view of human rights based on the international human rights instruments that the North has joined.

In addition, the North Korean authorities should reestablish their understanding of human rights at the domestic level in a fundamental way. In order to change the North's concept of human rights, the following preconditions must be in place. First, the North Korean authorities must abandon *Juche* and the military-first ideology as their guiding principles. This first requires the deletion of articles on *Juche* and the military-first ideology from the Socialist Constitution of the DPRK. Along with the removal of such principles, the North Korean authorities should no longer hold on to their habit of seeing human rights issues from the perspective of *Juche*. Second, North Korea must not hold on to collectivism, through which it has been interpreting human rights issues. This requires the deletion of the extreme collectivistic principle, "one for all and all for one," from the Socialist Constitution. Currently all North Korean people are required to live under organizations. However, such organizational and unified life must be gradually forgone. In order for the prosperity strategy through the economy-first principle to be successful, the authorities must flexibly manage the ego-centered thoughts that will inevitably emerge in the process of economic activities. Also, the authorities should not approach human rights issues as a matter of dispensation. Rather, the authorities must see it from the eyes of the North Korean people.

The survival strategy befitting the international political reality of the twenty-first century requires the North Korean authorities to replace "rule by man" with "rule of law" in regard to the human rights issue. The authorities should choose to raise the quality of socialist laws, which will help the regime's own safety. Specifically, two aspects of "rule of law" require special attention for the regime's security, which will earn the cooperation of the international community and legitimacy bestowed by the people. First, the North Korean authorities have to narrow the gap between their written laws and how the laws are actually executed. Only when domestic laws are executed properly can the authorities erase their designation as a violator of human rights. Priority must be given to properly abiding by North Korean criminal law and the criminal procedure code. Then, the authorities should get rid of notorious illegal accommodation facilities and political prisoners' camps. For the crimes that are defined in North Korea's criminal law as treason, the authorities should follow all of the procedures, investigation, preliminaries, prosecution, and trials as stated in the criminal procedure code. With the absence of political prisoners' camps, those who are convicted and sentenced for political crimes could be sent to labor camps. Meanwhile, criminals' families should not be subjected to *Yeon-jwa-je* (guilt by association).

Second, the North Korean authorities must end the public execution system, which has been criticized internationally for its inhumanity. Even in cases where the death penalty is imposed, it must be applied strictly according to criminal law, with a trial before sentencing, as stated in the criminal procedure code. In particular, as the law on enforcement of court decisions dictates, execution of the death penalty requires a decision of the Supreme People's Assembly. Also, it is appropriate to cancel the article on public trials. Punishment of the convicted must be dealt with from a human rights perspective; it must not be used for the purpose of evoking fear from the crowd. Furthermore, in order to raise the quality of "rule of law" and invigorate the economic activities of individuals, the deportation system must be abolished, and more efforts to narrow the gap between the law and actual practice are necessary. Although the central authority reorganizes the laws, if people who execute the law lack awareness of human rights, human rights crimes will remain at the root of the society. For the better execution of the law, the authorities should strengthen the capability of the state. Operating various educational programs for the enforcement units would be one way to accomplish this, and for this, the authorities need to cooperate with the international community. The expansion of technical cooperation with UNOHCHR is especially important. There are ways to build strong security for the regime by strengthening socialist laws, and the authorities must realize this.

Then, the North Korean authorities should gain international cooperation by accepting international human rights law and international protocols and reorganizing North Korean laws accordingly. In order to enhance the quality of the law, North Korea would be required to become a member of international conventions of human rights that it has not yet joined, such as the Convention against Torture. Next, the authority must modify the current law reflecting international human rights law. As a part of the effort, the authority must actively engage with UNOHCHR and the EU to cooperate in the area of human rights.

North Korea should move beyond the strong and prosperous state that it targeted in the nineteenth and twentieth centuries. It needs a strategy to secure the regime through the people-first policy and the acceptance of human rights. When the economy-first and the people-first approaches are combined in policy making, the safety of the regime can be better served. Rather than simply targeting security of the regime through the economy-first strategy, such as the development of agriculture and light industry, integrating people-first policies can be much more effective. In regard to the integration of the economy-first and the people-first approaches, at least a minimum level of citizen rights must be allowed. That includes people's right to move around in search of means to survive during the transition period to the market economy. In addition, the authorities should gradually widen the scope of reforms and institute an open-door policy for the successful integration of the economy-first and the people-first plans. For North Korea to succeed in the long run on its own, and prosper in the complex stage of the twenty-first century, it must also be able to compete in the stage of knowledge. Therefore, more books, especially books on market economy, need to be allowed to flow in with fewer limitations. In other words, the authorities should slowly allow the distribution of human rights–related books and ease its grip on information. No member nation of the EU completely blocks the activities of the press and media. Hence, in order to overcome the label "violator of human rights," the authorities ought to gradually open their doors to the press and media. This includes the use of mobile phones and computers, and especially Internet use among ordinary people as well as more freedom for the residents, including freedom to move around. Moreover, restrictions on foreign visitors need to be loosened. In particular, travel to rural areas has to be approved with fewer limitations. Also, contacts between foreign visitors and the North Korean people should be allowed. In the transition period from the planned economy to the market economy, people's increasing economic and social rights have to be considered when distributing resources. Especially for cooperation in development, the authorities have to work with the international community to make sure that cooperation occurs with the participation of the people in small divisions such as *li* and *dong*, or small regional divisions in Korea. Through such efforts, the authorities must guarantee people's basic participation in order to help people attain the capability to live on their own. In small regional divisions, the authorities must allow the people to participate in various areas such as development cooperation, policy decisions and execution, and evaluations. Considering the changes in social relations that inevitably follow the

Table 1. Development Stages for the Improvement of North Korea's Human Rights

	Human Rights and Security of the Regime	Approach to Human Rights	Acceptance Strategy	Mechanism of Human Rights Protection
Short term	Regime's security through *Juche*, military-first → consideration for economy-first	Radical relativism → strong relativism	Coercion, denial (essence), tactical Concession (secondary) → expansion of tactical concession	Talk with UN and individual states Strengthening application of law
Medium term	Regime's security through economy-first and people-first	Strong relativism → weak relativism	Tactical concession → expansion of human rights law within the state	Preference toward a society that partially accepts monitoring by UN and international NGOs.
Long term	Regime's security through people-first	Weak relativism	Human rights law spreads deeper into the society	Full acceptance of the system that protects domestic human rights

expansion of a market economy, the authorities must abandon the exclusion and discriminatory rules based on family background.

In particular, the extreme class principles that are stated in DPRK law (the criminal law and the code of criminal procedure) must be deleted. The North Korean authorities must aim for the "rule of law," reorganize domestic laws, and strengthen the ability to execute them, while accepting outside monitoring to a large degree.

South Korea and International Community

It is recommended that the international community approach North Korea's human rights issues in a way to change North Korean authorities' concerns regarding their regime security. Led by South Korea, North Korea will be able to establish an atmosphere where its human rights issues are not considered in a perspective of system reductionism. Understanding the multidimensional factors that violate North Korean human rights and inducing policy in North Korea to change in order to resolve these factors are additional steps. For dispelling worries about the security of the regime and political dispute among its authorities, South Korea and the international community could form a multidirectional cooperative network on human rights issues with North Korea. Through these types of multidimensional cooperative networks, gradual improvement of goals and strategy for action can be discussed. That being said, rather than just one-way directional means whereby the outsiders set the goal of improvement, it is very important to foster an atmosphere of discussion, where conversation and communication are emphasized in order to come up with ideas for improving the direction of strategy. The starting point must be mutual understanding and flexible adjustment on the concept of human rights.

First, relations between the United Nations and North Korea, in terms of human rights, should be continuously adjusted. More concretely, the UN human rights regime should change its approach toward North Korean human rights issues. This can be done in a way so that the UN itself adjusts means of problem-solving toward North Korean human rights issues, which will eventually lead North Korea to convert its military-first strategy to an economy-first and civilian-first strategy. A strategy could be changed from pressuring North Korea to promoting an environment of conversation and cooperation through adopting the North Korea Human Rights Resolution. The UN Special Rapporteur that selectively targets North Korea must be abolished. This approach would be designed to strengthen cooperation between the North Korean regime and the UN Thematic Special Rapporteur that is not specifically targeted at a certain country. The focus will be more effective when the UN committees that supervise the international human rights treaties, induce North Korea to adopt the international norm of human rights. Active communication with North Korean authorities through the Universal Periodic Review of the Human Rights Council and the creation of an environment for activating technical cooperation on human rights between the UN and North Korea are important. The Office of the High Commissioner for Human Rights (OHCHR) should support North Korea so that it can establish National Human Rights Action Plans (NHRAP). In this way, the South Korean government will have to cooperate closely with the UNOHCHR so that they can create a cooperative environment to strengthen human rights inside North Korea.

Second, the network of cooperation on human rights between individual nations and the North Korean regime should be diffused. The environment of the human rights dialogue that will induce a change of perception and policy in North Korea toward human rights can be created by Western countries. The EU, which has relatively friendly relations with North Korea, will be the one who needs to work actively on resuming a dialogue with North Korea on the issue, and countries like Canada and Australia, who are interested in human rights issues, will have to push the dialogue ahead. Countries like the United States, Japan, and South Korea are the ones who need to abolish laws related to North Korea while they carry forward human rights policies toward North Korea by establishing North Korean human rights laws in order to create an external atmosphere for the dialogue. This will give the United States and Japan a stepping-stone to solve North Korean human rights issues in a cooperative way. The United States will have to work on not letting the human rights issue agenda be a source of conflict in order to ameliorate its relations with North Korea by admitting the differences in perceptions of human rights and supporting North Korea so that North Korea can gradually achieve goals. Cooperation through "a North Korean human rights consultative group" will also be important. In addition, the nations participating in this human rights dialogue with North Korea will have to be active in the area of cooperation on the individual national level.

Third, a multilateral human rights regime in which North Korea can participate should be pushed ahead, rather than focusing on an isolating plan targeted at North Korea. However, North Korea seems to have a bad perception of the Helsinki process. It seems to believe that information-distributing strategies that are based on the Helsinki Final Act would be one of the main factors in the collapse of socialism. There is therefore a great possibility that North Korea might oppose any human rights plan that claims to follow the Helsinki process. Thus, creating a regional human rights system and a positive atmosphere that takes into account regional factors in East Asia is important.

Fourth, an environment for forming a network of cooperation on human rights between South and North should be created. Although considering the level of openness and the adjustment of the human rights strategy of North Korea is important, the development of North Korea should be promoted gradually, led by South Korea. Through this, the North Korean authorities' policy that is designed to promote its civilians' economic rights will be supported. A right atmosphere that is created by the international community will be an important factor to persuade North Korea to be active in changing its policy and system. When South Korea looks at its North Korea policy, it should focus on supporting development and taking a major role in eradicating poverty. Even though the number of North Korean defectors to South Korea is increasing because of North Korea's temporary humanitarian crisis, South Korea should strengthen South-North cooperation in combating large-scale North Korean defections to the South. In addition, presuming an improvement in the level of South-North relations, the National Security Law will have to be abolished. In this atmosphere, a South-North dialogue on human rights can be induced with the participation of civilians, scholars, and administration officials. In this dialogue, the subject of supporting human rights education programs and abolishing detention centers will be given priority, although the subject of concrete areas of cooperation will also be discussed.

Transformation Stage

North Korea

In order to avoid becoming a dropout state and to evolve into an intricate nation of the twenty-first century, what essentially North Korea should do is to take a transitional step for its regime so that it can transform to an open, norm-friendly regime. It should completely break away from seeing human rights issues from the perspective of the security of the regime: the transition should be developed in a way to gain legitimacy for the regime through an increase in civilian rights, and in a way to take a step from the level of acquiring a position within international norms to a level where it coincides with the international norms while claiming the process of five ascending steps. Carrying forward a strategy of internalization of international human rights norms will be helpful to achieve this. Additionally, joining in the major international covenants for human rights, observing and adhering to their laws completely, such as criminal law and the criminal procedure code, to make them accord with the international covenants for human rights, are also necessary. Introduction or reorganization of human rights–related legislation will be achieved by applying various international models/examples. Along with these steps, North Korea will have to abolish its class principle and collectivism that are prescribed in the domestic legislation related to human rights.

Breaking from a politically dominant society requires effort by North Korea to establish a stable law of socialism. In the human rights reports of the United Nations human rights body and international human rights NGOs, usually an overall poor score for North Korea's human rights is seen. A gap between North Korean legislation and actual practice could be minimized through reorganizing its entire domestic capability. For the legislative sector, institutionalizing education on human rights for legislators and ensuring the independence of the judicial branch should be the way to reorganize the legal system. Through this process, reorganization of the clause of any law that hinders the independence of the judicial branch and abolition of several similar jurisdictions, such as the Guidance Committee for Socialist Law, is especially significant. Modifying the institution of training prosecutors, judges, and attorneys is necessary for reforming legal institutions to institute the right of defense. Moreover, several other steps are also necessary to guarantee human rights, such as abolishing the system of labor camps and many other detention facilities, supplying books related to human rights, and introducing a human rights education system targeted at North Korean civilians and students.

North Korea is going to have to open its society to the international community. It will be possible for the North to make tangible progress when it guarantees its people's freedom of movement and their right to use information sourced from the outside. Permitting its citizens' use of the Internet and cell phones, and accepting major principles and standards regarding developmental cooperation, which are popularly used internationally, will be concrete steps to take. As the participating development and congregated approach of development and human rights are preferred in today's international community, North Korea also has to guarantee its people's full participation and rights while accommodating process.

North Korea also needs to work on being active in monitoring not only domestically but also internationally the issues of human rights. Achieving recognition from the international community on the existence of a North Korean intermonitoring system on human rights and pursuing cooperation with a monitoring system of UN human rights bodies, such as a UN Special Rapporteur, will be important in the process. Moreover, establishing an independent national human rights body is an effective way to deal with North Korea's human rights issues.

South Korea and the International Community

Institutionalizing South-North cooperation on human rights issues should proceed in a way that will enhance human rights on the Korean Peninsula, rather than just raising North Korean human rights issues in a one-way direction. It will be effective when the two countries work on forming a substantial opportunity to have a meaningful dialogue on human rights and discuss plans to enhance human rights of each country in various fields. In the process, South Korea can transfer its experience and lessons in a way to foster human rights awareness inside North Korea through inducing communication among the levels of labor, women, and youth. For example, cooperation between South and North can solve the problems caused by the belittlement of women in North Korean society.

North Korea cannot avoid modifying various laws and regulations in order to push ahead with the reform and open policy. This process for North Korea with South Korea's support will successfully establish the law of socialism inside of the country. South Korea is playing an important role in leading the situation in order to form not only South-North cooperation but also various legal cooperative channels with international organizations or other individual countries. For instance, South Korea can provide leadership in a body tentatively named "International Consultative Group for Supporting North Korean Rule of Law." This consultative group requires establishing a plan to support North Korean legal institutions and to consult with North Korean authorities. In this step, strengthening cooperation among the domestic departments that are related to North Korea, such as the National Human Rights Committee, or the Ministry of Justice, will support this kind of consultative group. Supporting an educational manual on legislative human resources inside South Korea can be one of the examples.

The international community needs to constantly work on not only establishing an East Asian regional system on human rights but also persuading and supporting the North Korean authorities in establishing their own independent institution of human rights. As was observed in the UN system of regular review of the human rights committee, the East Asian regional regimes on human rights will have to come up with a "Universal Periodic Review," which will initiate an equal discussion concerning North Korean human rights issues and those of neighboring countries. ■

■ References

Heo, Man-ho. 2008. "'Na-seon-hyeong 5dan-gye-lon'eu-lo bon bug-han-ui in-gwon-jeongchaeg: Helsinki Process ui jeog-yong" [North Korea's Human Rights Policy: Application of Helsinki Process]. *Gug-bang-yeon-gu* [National Defense Studies]. 51, 1.

Donnelly Jack. Jeong-won Park Trans. 2002. *International Human Rights*. Seoul: Oreum.

Jeon, Deokseong. 2004. *Seon-gun-jeongchi-e dae-han li-hae* [Understanding Military-first Politics]. Pyongyang: Pyongyang Publication.

Kim, Il-sung. 1996. "Answers to the Questions Raised by the Reporters from the *Washington Times*, the U.S" *Kim, Il-sung jeo-jag-jib 44* [Kim, Il-sung's Writing 44]. Pyongyang: Joseon Labor Party Publication.

Kim, Jong-il. 1998. "Interruptions to Socialism Cannot Be Allowed." *Kim, Jong-il Collection 13* [Kim, Jong-il seon-jib 13]. Pyongyang: Joseon Labor Party Publication.

Kim, Jong-il. 2000. "Protecting Identity and Nationality in Revolution: June·19, 1997." *Kim, Jong-il Collection 14*. [Kim, Jong-il seon-jib 14]. Pyongyang: Joseon Labor Party Publication.

Kim, Soo-Am. 2007. *Min-ju-ju-ui-wa in-gwon-e dae-han bug-han-ui in-sig-gwa dae-eung* [North Korea's Perception and Countermeasure for Democracy and Human Rights]. Seoul: Korea Institute for National Unification.

Lee, Keumsoon, and Soo-Am Kim 2009. *Bug-han-in-gwon chimhae gu-jo mich gae-seon-jeon-lyag* [North Korea's Human Rights Violations and Improvement Strategy]. Seoul: Korea Institute for National Unification.

Lee, Keumsoon, et al. 2008. *Gug-je gae-bal-i-lon hyeon-hwang* [Current Theories on International Development]. Seoul: Korea Institute for National Unification.

————. 2009. *Bug-han-in-gwon-baeg-seo 2009* [White Paper on North Korea's Human Rights 2009]. Seoul: Korea Institute for National Unification.

Lee, Nam-ju. 2008. "Junggug-e-seo in-gwon-gyu-beom-ui hwag-san-gwa han-gye" [Expansion and Limitation of Human Rights Norms in China]. presented at the 10th Korea Peace Forum. Seoul, August 29.

Risse, Thomas, and Kathryn Sikkink. 1999. "The Socialization of International Human Rights Norms into Domestic Practices: Introduction." In *The Power of Human Rights:International Norms and Domestic Change,* ed. Thomas Risse, Stephen C. Ropp, and Kathryn Sikkink. New York: Cambridge University Press.

Rodong Sinmun [Newspaper for Workers]. 1995. "Chamda-un in-gwon-eul ongho-ha-yeo" [Defending the Genuine Human Rights]. June 24.

———. 2006. "Jo-seon-in-gwon-yeon-gu-hyeob-hoe dae-byeon-in damhwa" [Talk with the Spokesperson of the Human Rights Studies Association]. August 19.

———. 2007. "Je-gug-ju-ui-ui 〈in-gwon〉 gongse-leul dan-ho-hi jisbu-si-ja" [Let Us Crush the 〈Human Rights〉 Attacks of the Imperialism]. August 17.

———. 2007. "Ban-yeog-sa-jeog-in ji-bae-ju-ui-oe-gyo-jeongchaeg-mi-gug in-gwon-gongse" [Hegemonic Foreign Policy-the Human Rights Attacks of the United States]. September 13.

Seo, Bo Hyeog. 2007. *Bug-han-in-gwon: I-lon · Sil-je · Jeongchaeg* [North Korea's Human Rights: Theory, Practice and Policy]. Seoul: Hanul.

■ The Contributors

Young-Sun Ha
Seoul National University

Young-Sun Ha is a professor in the Department of International Relations at Seoul National University and chairman of Global Net 21 at East Asia Institute. Professor Ha received both his B.A. and M.A. from Seoul National University, and received his Ph.D. in international politics from University of Washington. He served as director of the Center for International Studies, Seoul National University; director of the American Studies Institute, Seoul National University; president of the Korea Peace Studies Association; and research fellow at that Center for International Studies at Princeton University and the Center for International Peace in Stockholm. Coauthor of *East Asian Community: Myth and Reality* (2008),*Transformation of World Politics* (2007), *Network Knowledge State* (2006), *North Korean Nuclear Crisis and Peace on the Korean Peninsula* (2006), *Korean-American Alliance: A Vision and a Roadmap* (2006), *Korea's Grand Strategy for a New Century: Weaving a Network State* (2006),*Korean Diplomatic History and the Study of International Politics* (2005), *New Perspective of Changing World* (2004), *100 Years Plan for the Korean Peninsula in the 21st Century* (2004), and *World Politics of Cyberspace* (2001).

Seongji Woo
Kyung Hee University

Seongji Woo is currently an associate professor at the College of International Studies, Kyung Hee University. He worked as an assistant professor at the Institute of Foreign Affairs and National Security, Ministry of Foreign Affairs and Trade, Republic of Korea from March 2003 to February 2005. He also worked as a research fellow for the Center for International Studies, Seoul National University from August 2002 to August 2003. Dr. Woo received his Ph.D. from Indiana University, Bloomington. His English articles include "The Park Chung-hee Administration amid Inter-Korean Reconciliation in the Detente Period: Changes in the Threat Perception, Regime Characteristics, and the Distribution of Power" (Summer 2009, *Korea Journal*), "Triangle Research and Understanding Northeast Asian Politics"(Summer 2003, *Asian Perspective*), "South Korea's Search for a Unification Strategy" (Summer 2003, *Orbis*), and "Adversarial Engagement and Alliance Relations: Triangular Politics on the Korean Peninsula, 1988–94" (March 2001, *Issues & Studies*). His research interests are North Korean politics, Inter-Korean relations and international relations theory. Currently his research focuses on the dynamics of inter-Korean reconciliation processes in the early 1970s, 1990s, and 2000s.

Chaesung Chun
Seoul National University

Chaesung Chun is a professor in the Department of Political Science and International Relations in Seoul National University. He is now a visiting professor at Keio University in Tokyo. He is also director of Asian Security Initiative of East Asia Institute. He is a member of the advisory committee for the Ministry of Foreign Affairs and Trade and the Ministry of Unification. He received his M.A. degree from Seoul National University and his Ph.D. degree from Northwestern University in the field of international relations theory. Major fields include international relations theory, security studies, South Korean foreign policy, and East Asian security relations. Major articles include "A Study on the Formation of European Modern States System," "Critique of Constructivism from the Perspective of Postmodernism and Realism," and "The Rise of New Powers and the Responding Strategies of Other Countries."

Jihwan Hwang
University of Seoul

Jihwan Hwang is an assistant professor at the University of Seoul, Korea. Professor Hwang was an assistant professor of North Korean Studies at Myongji University and a research fellow at the Institute for Unification Studies of Seoul National University. His research interests include East Asian international relations and security studies, focusing on the North Korean nuclear crisis. His main concern is how to explain North Korea's foreign policy in terms of international relations theories. He has published numerous articles, and his recent publications include "International Relations Theory and the North Korean Nuclear Crisis," "Offensive Realism, Weaker States, and Windows of Opportunity: The Soviet Union and North Korea in Comparative Perspective," "The Second Nuclear Crisis and U.S. Foreign Policy," and "Rethinking the East Asian Balance of Power." Professor Hwang received his Ph.D. in political science from the University of Colorado at Boulder.

Dongho Jo
Ewha Womans University

Dongho Jo is a professor of North Korean Studies at Ewha Womans University. Before joining the university in 2007, he had worked at the Korea Development Institute for 16 years as senior fellow and director of North Korean economic studies. He was graduated from Seoul National University and received his Ph.D. degree in economics from University of Pennsylvania in 1991. He has also had lots of professional experience. He is now policy advisor for the chief secretary to the president of Korea for diplomacy and national security, advisor for the national assembly budget office, advisor for the advisory council on democratic and peaceful unification for the president of Korea, advisor for promoting inter-Korean exchanges and cooperation at the Ministry of Unification, etc. His research areas are mainly the North Korean economy and inter-Korean cooperation.

Soo-Am Kim
Korea Institute for National Unification

Dr. Kim, Research Fellow at the Korea Institute for National Unification, received his doctoral degree at Seoul National University and currently serves as a standing committee member of the National Unification Advisory Council and a policy advisor at the Ministry of Foreign Affairs and Trade. His research is focused on North Korean human rights, aid to North Korea, and inter-Korean humanitarian agendas. Dr. Kim has published numerous research papers and briefs. Some of his most recent briefs are "Daebukjiwongwa gungminjeok habui" (Aid to North Korea and the National Consensus), "Yueningwollejimgwa bukhaningwon: "jeollyak"gwa "gwangye"reul jungsimeuro"(The UN Human Rights Regime and North Korean Human Rights: With "Strategy" and "Relations" at the Center), and "Helsingki choejonguijeongseoui ui uiwa teukjing: ingwonuijereul jungsimeuro" (Meaning and Features of Helsinki's Final Act: With the Human Rights Agenda at the Center).